FUNDAMENTALS OF PHYSICS

Alessio Mangoni, PhD

©2020 Alessio Mangoni. All rights reserved.
ISBN: 9798655711945

DR. ALESSIO MANGONI, PHD

Scientist and theoretical particle physicist, researcher on high energy physics and nuclear physics, author of many scientific articles published on international research journals, available at the link:

http://inspirehep.net/author/profile/A.Mangoni.1

https://www.alessiomangoni.it

I edition, June 2020

Contents

Contents 5

Introduction 15

I Quantum Mechanics

1 Introduction 19

2 The wave function 23

3 The Schrödinger equation 27
3.1 Free particle equation   28
3.2 General equation   33

3.3	Continuity equation	35
4	**Wave packets**	**37**
5	**Normalization**	**45**
6	**Fourier transform**	**51**
6.1	Interval of length 2π	51
6.2	Interval of length L	53
6.3	Infinite interval	55
6.4	Coordinate and momentum space	56
7	**Expectation value**	**63**
8	**Operators**	**67**
8.1	Position operator	68
8.2	Momentum operator	69
8.3	Energy operator	72
8.4	Angular momentum operator	74
8.5	Spherical coordinates	76

9	Commutation relations	81
10	Uncertainty principle	87
11	Eigenvalue equations	93
11.1	Position operator	94
11.2	Momentum operator	96
11.3	The operator $\hat{L}_z$	98

II Particle Physics

12	Introduction	105
13	Natural units	107
14	Bases of relativity	109
14.1	Four-vectors	109
14.2	Lorentz transformations	111
14.3	Relativistic kinematics	113
14.4	Invariant mass	115

15 Particles 117

15.1 Elementary particles 118
15.1.1 Quarks 118
15.1.2 Leptons 118
15.1.3 Quark model 119

15.2 Fundamental interactions 120

15.3 Hadrons 123
15.3.1 Mesons 123
15.3.2 The Yukawa meson 125
15.3.3 Baryons 127

15.4 Nucleons 127

15.5 Cosmic rays 129

15.6 The pion 130

15.7 The muon 131

15.8 Particles with strangeness 131
15.8.1 Kaons 133
15.8.2 Hyperons 133

16 Energy loss 135

16.1 Ionization energy loss 135

16.2	Electron energy loss	136
16.3	Photon energy loss	137
16.4	Hadron energy loss	137

17 Quantum numbers and symmetries 139

17.1	The strangeness	139
17.2	The parity	140
17.3	Parity of the photon	141
17.4	Parity of a two-particle system	141
17.5	Charge conjugation	142
17.6	Charge conjugation of the photon	142
17.7	Charge conjugation of the pion	143
17.8	Time reversal	143
17.9	CPT theorem	143
17.10	Baryon number	144
17.11	Lepton number	144
17.12	Isospin	145
17.13	Hypercharge	146
17.14	The Gell-Mann-Nishijima formula	146

17.15	*G*-parity	147
17.16	Helicity	148
17.17	Chirality	148

18 Scattering and decays 151

18.1	Reference frames	151
18.2	The invariant quantity *s*	152
18.3	Mandelstam variables	154
18.4	Two-body elastic scattering	155
18.5	Fermi's golden rule	159
18.6	Cross section	160
18.6.1	Beam intensity reduction	163
18.6.2	Luminosity	165
18.6.3	Two-body cross section	165
18.7	Decays	166

III Theoretical Physics

19 Introduction 171

20 Lagrangian and Hamiltonian 173
- 20.1 Lagrangian field theory — 173
- 20.2 Hamiltonian field theory — 175

21 Symmetries and gauge invariance — 177
- 21.1 Symmetries and conservation laws — 177
- 21.2 Gauge invariance — 179

22 The Klein-Gordon field 185
- 22.1 Klein-Gordon equation — 185
- 22.2 Klein-Gordon Lagrangian — 186
- 22.3 Klein-Gordon Hamiltonian — 188

23 The electromagnetic field 189
- 23.1 Maxwell's equations — 189
- 23.2 Gauge invariance — 191
- 23.3 Maxwell Lagrangian — 193

24 The Dirac field 197
- 24.1 Dirac equation — 197

24.2	Properties of γ matrices	198
24.3	Dirac Lagrangian	203
24.4	Dirac Hamiltonian	203
24.5	Free particle solutions	206

25 Quantum electrodynamics 215

25.1	Interaction Lagrangian	215
25.2	Interaction Hamiltonian	218
25.3	Field operators	218
25.4	The S matrix	220

IV Condensed Matter Physics

26 Introduction 227

27 Brownian motion and diffusion ... 229

27.1	Introduction	229
27.2	Einstein relation	230
27.3	Fick's laws	233

27.4	Random walker	234
27.5	Langevin equation	236
27.6	Fokker-Planck equation	241
27.7	Boltzmann equation	242

28 Drude model 245

28.1	Introduction	245
28.2	Electric conductivity	246
28.3	Hall effect	248
28.4	Thermal conductivity	250
28.5	Seebeck effect	252

29 Sommerfeld model 255

29.1	Quantum treatment	255
29.2	Internal energy	257
29.3	Sommerfeld expansion	263

30 Mechanical properties of solids .. 269

| 30.1 | Introduction | 269 |
| 30.2 | Young's modulus | 270 |

30.3	Poisson's ratio	270

31 Lattice defects ... 271

31.1	Introduction	271
31.2	Point defects	272
31.3	Color centers	275
31.4	Dislocations	276

32 Semiconductors ... 277

32.1	Intrinsic semiconductor	277
32.2	Extrinsic semiconductor	280

Introduction

This book aims to provide solid bases for the study of physics for the university and it is divided into four parts, each dedicated to a fundamental branch of physics: quantum mechanics, theoretical physics, particle physics and condensed matter physics. In the first part we start with the concept of wave function, until the Heisenberg uncertainty principle. In the second part, after recalling the basic concepts of relativity, we treat the elementary particles and the hadrons, arriving to the notions of scattering and cross section. The third part is dedicated to the theoretical physics, where we analyze the field theory and the concepts of Lagrangian and Hamiltonian, introducing the quantum electrodynamics (QED), passing through the Klein-Gordon, Dirac and Maxwell fields. In the last part of the book we expose the basics of the condensed matter physics, including diffusion and Brownian motion, Drude and Sommerfeld models, the calculation of specific heat and the principal mechanical properties of solids, with references to lattice defects and semiconductors.

Part I

Quantum Mechanics

Chapter 1

Introduction

In this first part we will provide a rigorous, but intuitive and therefore suitable for most, theoretical introduction of non-relativistic quantum mechanics. This theory describes systems of particles of atomic scale dimensions, but with small velocity compared to the speed of light in vacuum, for which the relativistic effects can be neglected. There are four fundamental forces in nature: the nuclear strong force, the electromagnetic force, the nuclear weak force and the gravitational one. The two theories that should be considered for a modern description of nature are the Einstein's special relativity and the quantum mechanics. Nowadays all the fundamentals interactions except the gravitation are described by quantum theories of fields (relativistic theories) such as the quantum electrodynamics (QED) and the quantum chromodynamics (QCD). For this reason the study of quantum mechanics represents a fundamental objective. In this part we will treat only the non-relativistic quantum mechanics which represents also the basis for its relativistic formulation (which is often formulated through field theories). In this part we will cover the following topics:

- the wave function;
- the Schrödinger equation (free particle, general equation and continuity equation);
- the wave packets;
- the normalization;
- complete systems and Fourier transform;
- coordinate and momentum space;
- the expectation value;
- the operators (position, momentum, energy, angular momentum);
- the operators in spherical coordinates;
- the commutation relations;

- the eigenvalue equations;
- the Heisenberg uncertainty principle.

Chapter 2

The wave function

Let's start by saying that the description of a quantum system occurs through a function, called wave function, associated to the system. This is a function of time and space (x, y, z coordinates) and, in general, it is a complex number. It is usually denoted by the Greek letter $\Psi(x, y, z, t)$ and must satisfy some properties which we will list shortly. First of all, the formulation of quantum mechanics is based on the so-called "Copenhagen interpretation" and asserts that everything that can be known about a system is contained in its wave function. In particular, the probability of finding the system in the volume element between (x, y, z) and $(x + dx, y + dy, z + dz)$

at a certain moment t is given by

$$|\Psi(\vec{x},t)|^2 d^3x.$$

Note that it is a non-negative real number being the square modulus of a complex number. If we integrate the probability of finding a system on all the available volume we should obtain 1 (which corresponds to a percentage of 100%), that is, the certainty of finding it somewhere on the available volume. As we will see later, this cannot happen for a free particle since, also intuitively, the probability density of finding it somewhere is constant and if we integrate a constant on an infinite volume we will find infinite and not 1. The solution is to limit the available volume of the particle, in fact also in nature it can never be infinite. Such a normalization is called "box normalization" and will be discussed later. When the integral of the square modulus of the wave function, extended to the available volume, is 1 then it is said that the wave function is normalized to 1 and its square modulus gives the probability density of the particle presence. For normalized wave functions it therefore happens that

$$\int |\Psi(\vec{x},t)|^2 d^3x = 1.$$

We now list the physical requirements that a wave function must satisfy in order to describe a quantum system:

1. the wave function must be everywhere continuous. Being connected with the probability of finding a particle in a volume in a certain time it cannot be discontinuous, otherwise there would be different probabilities depending on the way of calculating the volume;

2. the wave function must be limited everywhere. In fact, it makes no sense to speak of infinite probability of finding the system somewhere (the maximum probability is 1);

3. the wave function must be a single valued function, i.e. monodromic. In fact, you cannot have more probabilities for a given point and a given time.

To conclude this chapter on the wave function of a quantum system (or for a particle, in general) we illustrate the so-called superposition principle. Meanwhile, let's say that two wave functions that differ in the normalization constant or in a generic multiplicative complex constant describe the same system. In addition, given two wave functions that describe the same system then a linear combination of them will also describe that system. For practical purposes and for the concept of probability given to the square modulus of the wave function we will always choose a wave function normalized to 1 (when possible, for example for free particle we will adopt the so-called "box normalization", as we will see later). We can multiply a normalized wave function by a phase factor of the type $e^{i\alpha}$ with modulus 1. In general if $\Psi(x,y,z,t)$ is the normalized wave function for a system then also

$$\Psi(\vec{x},t)\,e^{i\alpha},$$

with α an arbitrary real constant, it will be a normalized wave function for the same system since

$$|\Psi(\vec{x},t)\,e^{i\alpha}|^2\,d^3x = |\Psi(\vec{x},t)|^2\,d^3x,$$

because

$$|e^{i\alpha}|^2 = e^{i\alpha}e^{-i\alpha} = 1, \quad \alpha \in \mathbb{R}.$$

Chapter 3

The Schrödinger equation

We now come to the equation on which all non-relativistic quantum mechanics is based. This is a partial differential equation called Schrödinger equation, from the name of the scientist who formulated it for the first time. The essential problem is to find the wave function for a certain quantum system, in fact, once found, we can access the calculation of all probabilities and physical observables, using its square modulus, as said in the previous chapter. The Schrödinger equation has this purpose, once solved it provides the wave function of the system. The real problem is to solve it, in the sense that not all systems are described by a Schrödinger equation that admits analytical solutions, in terms of elementary functions and in these cases you can proceed only with numerical calculations or with approximate methods, using the so-called perturbation theory. Firstly we present the Schrödinger equation for the simplest case of a free particle and then we will show its general form.

3.1 Free particle equation

The Schrödinger equation for a free particle of mass m is

$$i\hbar \frac{\partial \Psi(\vec{x},t)}{\partial t} = -\frac{\hbar^2 \vec{\nabla}^2 \Psi(\vec{x},t)}{2m},$$

where i is the imaginary unit, with

$$i^2 = -1,$$

the symbol $\hbar$ is called the reduced Planck constant, given by

$$\hbar = \frac{h}{2\pi}$$

3.1 Free particle equation

and $\vec{\nabla}^2$, where ∇ is the nabla operator, is said Laplacian (Laplace operator), a differential operator given by

$$\vec{\nabla}^2 = \vec{\nabla} \cdot \vec{\nabla} = \frac{\partial^2}{\partial x^2} + \frac{\partial^2}{\partial y^2} + \frac{\partial^2}{\partial z^2},$$

in Cartesian coordinates, with

$$\vec{\nabla} = \left(\frac{\partial}{\partial x}, \frac{\partial}{\partial y}, \frac{\partial}{\partial z}\right).$$

Let's focus first on the one-dimensional case, the equation becomes

$$i\hbar \frac{\partial \Psi(x,t)}{\partial t} = -\frac{\hbar^2}{2m} \frac{\partial^2 \Psi(x,t)}{\partial x^2}.$$

This equation can be solved by separation of variables assuming that

$$\Psi(x,t) = \psi(x)\phi(t),$$

from which we obtain

$$i\hbar \psi(x) \frac{\partial \phi(t)}{\partial t} = -\frac{\hbar^2}{2m} \phi(t) \frac{\partial^2 \psi(x)}{\partial x^2}$$

and, dividing both members by the wave function,

$$i\hbar \frac{\psi(x)}{\Psi(x,t)} \frac{\partial \phi(t)}{\partial t} = -\frac{\hbar^2}{2m} \frac{\phi(t)}{\Psi(x,t)} \frac{\partial^2 \psi(x)}{\partial x^2},$$

which is identical to

$$i\hbar \frac{1}{\phi(t)} \frac{\partial \phi(t)}{\partial t} = -\frac{\hbar^2}{2m} \frac{1}{\psi(x)} \frac{\partial^2 \psi(x)}{\partial x^2}.$$

This equation has at first member a quantity which depends only on the variable t and at second member a quantity which depends only on x. The only way these two quantities are equal is that they are both constant. Said E this constant (it will be identified as the energy of the free particle described by the wave function) we set

$$i\hbar \frac{1}{\phi(t)} \frac{\partial \phi(t)}{\partial t} = E$$

and

$$-\frac{\hbar^2}{2m} \frac{1}{\psi(x)} \frac{\partial^2 \psi(x)}{\partial x^2} = E.$$

The first equation is an ordinary differential equation and has as the solution

$$\phi(t) = c_1 e^{-iEt/\hbar},$$

3.1 Free particle equation

with c_1 constant, while the second has the generic solution

$$\psi(x) = c_2 e^{ikx} + c_3 e^{-ikx},$$

with c_2 and c_3 constants and

$$k = \sqrt{2mE}/\hbar$$

which will be identified as the module of the wave vector of the particle. The more general solution can therefore be written as

$$\Psi(x) = C_1 e^{i(kx-\omega t)} + C_2 e^{-i(kx+\omega t)},$$

with C_1 and C_2 constants and

$$E = \hbar\omega.$$

The time independent equation, that for example depends only on the variable x, is called the stationary Schrödinger equation. In general we will try to solve this equation knowing that the time evolution is given by the addition of a phase

factor of the type
$$e^{-iEt/\hbar},$$
unless you have a problem where the particle is subject to a potential that explicitly depends on time.

Concerning the stationary equation we obtain that the wave function for a free particle (with defined wave vector, with modulus k) is a plane wave of the type

$$e^{ikx},$$

where k is connected to the energy E and to the momentum p by
$$E = \frac{p^2}{2m} = \frac{\hbar^2 k^2}{2m}, \quad p = \hbar k.$$

In the three-dimensional case, in a similar way, the solutions are of the type
$$\psi(\vec{x}) = e^{i\vec{k}\cdot\vec{x}}$$

and similarly

$$E = \frac{\vec{p}^{\,2}}{2m} = \frac{\hbar^2 \vec{k}^2}{2m}, \quad \vec{p} = \hbar \vec{k},$$

with the wave vector $\vec{k}$ whose modulus is the wave number k, as already said.

3.2 General equation

The general case occurs when the system (or the particle) is subject to a certain potential (we strictly mean a potential energy, which we will however call potential, letting the reader understand, depending on the situation, which physical quantity it refers to). We call this potential with the symbol V and the Schrödinger equation becomes

$$i\hbar \frac{\partial \Psi(\vec{x},t)}{\partial t} = -\frac{\hbar^2 \vec{\nabla}^2 \Psi(\vec{x},t)}{2m} + V(\vec{x},t)\Psi(\vec{x},t).$$

Typically we are dealing with potentials not dependent on time and we are interested in the stationary Schrödinger equation that takes the form

$$-\frac{\hbar^2 \vec{\nabla}^2 \psi(\vec{x})}{2m} + V(\vec{x})\psi(\vec{x}) = E\psi(\vec{x}),$$

or, in the one-dimensional case,

$$-\frac{\hbar^2}{2m}\frac{\partial^2 \psi(x)}{\partial x^2} + V(x)\psi(x) = E\psi(x).$$

We anticipate that the stationary Schrödinger equation can be written also as an eigenvalue equation as

$$\hat{H}\psi = E\psi,$$

where $\hat{H}$ is the Hamiltonian (differential) operator which is composed of the kinetic energy operator $\hat{T}$ (which is a differential operator) added to the potential operator $\hat{V}$ (which is a multiplicative operator)

$$\hat{H} = \hat{T} + \hat{V}.$$

The kinetic energy operator is given by

$$\hat{T} = \frac{\hat{p}^2}{2m},$$

where $\hat{p}$ denotes the momentum operator that has the form

$$\hat{p} = -i\hbar\vec{\nabla},$$

while

$$\hat{V} = V.$$

3.3 Continuity equation

Returning to the Schrödinger equation that can be written as

$$i\hbar\frac{\partial \Psi}{\partial t} = -\frac{\hbar^2 \vec{\nabla}^2 \Psi}{2m} + V\Psi,$$

with E the energy and V the potential to which the particle is subjected, we are interested in calculating the conjugate complex of the equation. Since the potential V is real we can write

$$-i\hbar\frac{\partial \Psi^*}{\partial t} = -\frac{\hbar^2 \vec{\nabla}^2 \Psi^*}{2m} + V\Psi^*,$$

multiplying the former equation by the conjugate complex of the wave function (ψ^*), the latter equation by the wave function (ψ) and subtracting from each other, we obtain

$$i\hbar\left(\Psi^*\frac{\partial \Psi}{\partial t} + \Psi\frac{\partial \Psi^*}{\partial t}\right) = -\frac{\hbar^2}{2m}\left(\Psi^*\vec{\nabla}^2\Psi - \Psi\vec{\nabla}^2\Psi^*\right),$$

or also

$$i\hbar\frac{\partial}{\partial t}\left(\Psi^*\Psi\right) = -\frac{\hbar^2}{2m}\vec{\nabla}\left(\Psi^*\vec{\nabla}\Psi - \Psi\vec{\nabla}\Psi^*\right).$$

Hence, having defined the square modulus of the normalized wave function as the probability density of presence that we call with the letter ρ,

$$\frac{\partial \rho}{\partial t} = \frac{i\hbar}{2m} \vec{\nabla} \left(\Psi^* \vec{\nabla} \Psi - \Psi \vec{\nabla} \Psi^* \right),$$

with

$$\rho = |\Psi|^2 = \Psi \Psi^*.$$

The continuity equation is thus obtained

$$\frac{\partial \rho}{\partial t} + \vec{\nabla} \cdot \vec{J} = 0,$$

with the probability current density given by

$$\vec{J} = -\frac{i\hbar}{2m} \left(\Psi^* \vec{\nabla} \Psi - \Psi \vec{\nabla} \Psi^* \right).$$

Chapter 4

Wave packets

We have seen that the general solution of the Schrödinger equation for a free particle (i.e. where $V = 0$) is given by an overlap of plane waves, each with a certain wave number k. A free particle with a well-defined k (or, equivalently, a well-defined momentum p) has the following wave function

$$\psi(\vec{x}) = e^{i\vec{k}\cdot\vec{x}}$$

and the particle cannot be located in space since the position probability density (square modulus of the wave function) is constant and therefore it is "everywhere with equal probability". If we want a particle that is spatially localized we can build the so-called "wave packet", given by the superposition of several plane waves with different wave numbers k and appropriate "weight" coefficients. In formula we can write

$$\Psi(\vec{x},t) = \int d^3k\, C(\vec{k}) e^{i(\vec{k}\cdot\vec{x}-\omega t)}.$$

This wave function can describe a possible state of a free particle since it represents a solution of its Schrödinger equation. The weight function $C(k)$ is called the spectral function and if it is a square summable function (i.e. the integral of its square modulus does not diverge) then it is also the wave function (therefore the integral of the probability density does not diverge and it makes sense to speak of presence probability). Now consider the one-dimensional case and suppose that we have a state given by the superposition of plane waves with wave numbers k between the following values

$$\tilde{k} - b \leq k \leq \tilde{k} + b.$$

As written above, the wave function will then be

$$\Psi(x,t) = \int_{\tilde{k}-b}^{\tilde{k}+b} dk\, C(k) e^{i(kx-\omega t)}.$$

Suppose that b is small number and that $C(k)$ can be considered constant (slowly variable in the interval in which the integration is carried out), therefore

$$\Psi(x,t) = C \int_{\tilde{k}-b}^{\tilde{k}+b} dk\, e^{i(kx-\omega t)}.$$

We can expand ω in powers of

$$k - \tilde{k},$$

knowing that

$$\omega = \frac{E}{\hbar} = \frac{\hbar k^2}{2m}.$$

Then

$$\omega(k) = \omega(\tilde{k}) + (k - \tilde{k}) \frac{d\omega}{dk}\bigg|_{k=\tilde{k}},$$

with

$$\frac{d\omega}{dk}\bigg|_{k=\tilde{k}} = \frac{\hbar \tilde{k}}{m}.$$

By replacing
$$y = k - \tilde{k}, \quad dy = dk,$$
the integral that gives the wave function becomes
$$\Psi(x,t) = Ce^{i(\tilde{k}x - \omega(\tilde{k})t)} \int_{-b}^{b} dy\, e^{i(x - t\partial\omega/\partial k|_{k=\tilde{k}})y}.$$

Calculating we obtain
$$\Psi(x,t) = 2Ce^{i(\tilde{k}x - \omega(\tilde{k})t)} \frac{\sin\left[(x - t\partial\omega/\partial k|_{k=\tilde{k}})b\right]}{x - t\partial\omega/\partial k|_{k=\tilde{k}}},$$

that is the product between a plane wave (associated with the central wave number of the packet) and an amplitude factor that modifies its shape. This function is also normalizable, being a square summable function. We write
$$\Psi(x,t) = 2Ce^{\tilde{k}x - \tilde{\omega}t} \frac{\sin\left[(x - \hbar\tilde{k}t/m)b\right]}{x - \hbar\tilde{k}t/m},$$

with
$$\tilde{\omega} = \omega(\tilde{k}).$$

Let's consider the amplitude

$$A(x,t) = 2C \frac{\sin\left[(x - \hbar \tilde{k} t/m)b\right]}{x - \hbar \tilde{k} t/m},$$

which has a maximum for $x = 0$, given by $2Cb$, and oscillates reducing its value as can also be seen from the qualitative plot shown in figure 4.0.1 (calculated at $t = 0$, since it does not change its shape in time, it only moves along the x axis, in the positive direction).

Figure 4.0.1

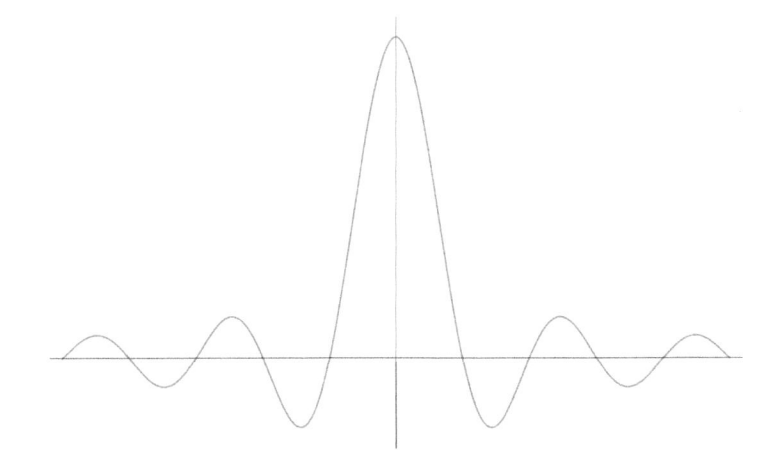

Furthermore, the square modulus of the wave function, i.e. the presence probability density, is significantly different from zero only in a restricted spatial region so we say that a wave packet describes a free particle located within a certain spatial region. A qualitative plot of the trend of the probability density is shown in figure 4.0.2.

Figure 4.0.2

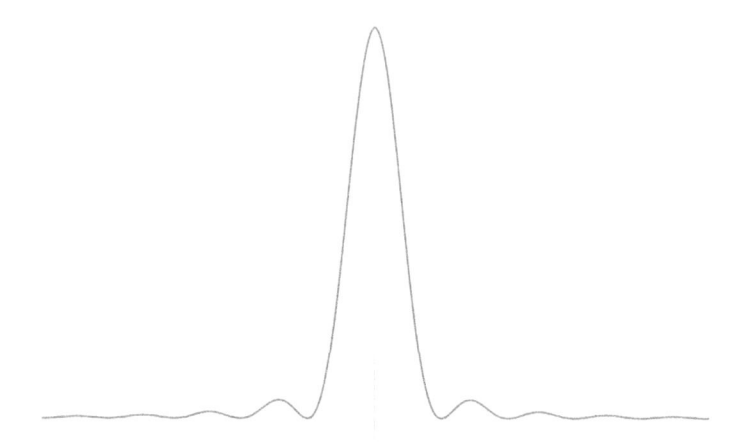

In order for the particle to be localized we had to combine several plane waves each with a different k (but all within b from a certain $\tilde{k}$). So we have a free particle with an "uncertainty" on its position, but also on its k (i.e. on its momentum, since they are mathematically connected through the reduced Planck constant, $\hbar$). The more the particle will be spatially localized the less it will be defined its momentum and, therefore, its velocity and vice versa. Note that in the extreme case in which a particle has a well defined k it is completely delocalized (single plane wave). This concepts is enclosed, mathematically, in the so-called Heisenberg uncertainty principle which will be discussed later. Returning to the wave packet, we can calculate its velocity (group velocity) along the x axis. This is given by

$$v_g = \frac{d\omega}{dk}\bigg|_{k=\tilde{k}} = \frac{\hbar \tilde{k}}{m} = \frac{\tilde{p}}{m},$$

which corresponds to the velocity of a particle with a momentum equal to the "average" momentum of the wave packet (since the regime is not relativistic, it is $p = mv$). Finally, for the normalization of the wave functions of the packet we can write

$$C = \frac{1}{2\sqrt{\pi b}}.$$

This can be calculated using the known integral

$$\int_{-\infty}^{+\infty} \frac{\sin^2 x}{x^2} dx = \pi.$$

Chapter 5

Normalization

We discussed previously the normalization of a wave function. Since its square modulus represents the presence probability density of a particle, we should impose that the probability of finding the particle somewhere is equal to 1. In general, if the function is a square summable function, we impose that the arbitrary multiplicative constant, called C, satisfies

$$\int d^3x |C\psi(\vec{x})|^2 = 1$$

and, usually, we choose C to be real with

$$C = \sqrt{\frac{1}{\int d^3x |\psi(\vec{x})|^2}}.$$

For a free particle with a well defined momentum (and, therefore, well defined k) the wave function (only the spatial part, since the temporal one, being a multiplicative phase with modulus 1, does not change the total squared modulus) is proportional to the plane wave

$$\psi(\vec{x}) = e^{i\vec{k}\cdot\vec{x}}.$$

This function cannot be normalized over the whole space, as anticipated. If the free particle has a defined momentum we can limit the available space where it can be. This assumption leads to the so-called box normalization, that is, it is assumed that the space available for the particle is represented, for example, by a cube with side L. In this case we consider the monochromatic plane wave (i.e. of defined momentum p or, similarly, of defined wave number k)

$$\psi(\vec{x}) = e^{i\vec{k}\cdot\vec{x}}$$

and we impose that, having fixed the origin of the Cartesian system x, y, z on one edge of the cube, the wave function assumes the same value on the faces of the cube, i.e.

$$\psi(x, y, z) = \psi(x+L, y, z)$$

and similarly for y and z. Being

$$e^{i\vec{k}\cdot\vec{x}} = e^{k_x x + k_y y + k_z z} = e^{k_x x} e^{k_y y} e^{k_z z},$$

we obtain

$$e^{k_x x} = e^{ik_x(x+L)},$$

or

$$e^{k_x L} = 1.$$

This implies that the angle of this phase is an integer number, multiple of 2π, i.e.

$$k_x L = 2\pi n, \quad n \in \mathbb{Z}$$

and similarly for y and z. Summarizing we have the following quantization conditions of the wave vector

$$\begin{cases} k_x = \frac{2\pi}{L} n_x \\ k_y = \frac{2\pi}{L} n_y \\ k_z = \frac{2\pi}{L} n_z \end{cases}, \quad n_x, n_y, n_z \in \mathbb{Z},$$

or

$$\vec{k} = \frac{2\pi}{L}(n_x, n_y, n_z), \quad n_x, n_y, n_z \in \mathbb{Z}.$$

Furthermore, the normalization of the wave function is calculated by setting the multiplicative constant C such that

$$\int_0^L dx \int_0^L dy \int_0^L dz \, |C\psi(\vec{x})|^2 = 1,$$

from which

$$|C|^2 = \frac{1}{L^3}$$

and we can then choose the real constant

$$C = \frac{1}{L^{3/2}}.$$

The wave function becomes

$$\psi_{\vec{k}}(\vec{x}) = \frac{1}{\sqrt{V}} e^{i\vec{k}\cdot\vec{x}}, \quad V = L^3,$$

with V the volume of the cube where the particle is confined and we recall that the wave vector $\vec{k}$ is quantized as seen above.

Chapter 6

Fourier transform

6.1 Interval of length 2π

The set of the following functions

$$\beta_n(x) = \frac{e^{inx}}{\sqrt{2\pi}}, \quad n \in \mathbb{Z},$$

represents a complete orthonormal system for the class of the square summable functions in any interval of amplitude 2 *pi* and, in particular, between $-\pi$ and π. For the elements of the orthonormal system we have the scalar product

$$\left(\beta_n(x), \beta_m(x)\right) = \int_{-\pi}^{\pi} \beta_n^*(x) \beta_m(x)\, dx = \delta_{nm},$$

where we have used the Kronecker delta, which assumes the value 1 when $n = m$ and 0 otherwise. Completeness also implies that

$$\sum_{n=-\infty}^{+\infty} \beta_n(x)\beta_n^*(x') = \frac{1}{2\pi} \sum_{n=-\infty}^{+\infty} e^{in(x-x')} = \delta(x-x'),$$

where in the last member we used the Dirac delta distribution. Any square summable function in the chosen range $(-\pi, \pi)$ can therefore be expressed as a Fourier series which converges to the function "almost everywhere" in the interval, i.e. except for some measure zero subset. The coefficients of the Fourier series for the function $f(x)$ are said to be the "components" of the function with respect to the orthonormal basis written above, i.e.

$$\left(\beta_n(x), f(x)\right) = \int_{-\pi}^{\pi} \beta_n^*(x) f(x) \, dx = \delta_{nm},$$

or

$$\left(\beta_n(x), f(x)\right) = \frac{1}{\sqrt{2\pi}} \int_{-\pi}^{\pi} e^{-inx} f(x) \, dx = \delta_{nm}.$$

Calling
$$a_n = \left(\beta_n(x), f(x)\right),$$
the Fourier series can be written as
$$f(x) = \sum_{n=-\infty}^{+\infty} a_n \beta_n(x) = \frac{1}{\sqrt{2\pi}} \sum_{n=-\infty}^{+\infty} a_n e^{inx}.$$

Furthermore, we have the Parseval equation
$$\sum_{n=-\infty}^{+\infty} |a_n|^2 = \int_{-\pi}^{\pi} |f(x)|^2 dx.$$

6.2 Interval of length L

In the case of an interval of amplitude different by 2π, the complete orthonormal system can be constructed as
$$\psi_n(x) = \frac{e^{ik_n x}}{\sqrt{L}}, \quad k_n = \frac{2\pi}{L} n, \quad n \in \mathbb{Z},$$
by the following replacement in the initial complete orthonormal system
$$x \to \frac{2\pi}{L} x.$$

We assigned the letter ψ to the functions of this complete orthonormal system since they are practically the box normalized free particle wave functions we have encountered previously. All the results shown above still apply, obviously changing the integration extremes with $-L/2$ and $L/2$, respectively, and using the correct normalization constant of the new orthonormal system. We remember that any square summable function between $-L/2$ and $L/2$

$$\psi(x) \in \mathcal{L}^2\left(-\frac{L}{2}, \frac{L}{2}\right),$$

can be expressed as a Fourier series "almost everywhere" (everywhere except some measure zero subset), i.e.

$$\psi(x) = \sum_n a_n \psi_n(x) = \frac{1}{\sqrt{L}} \sum_n a_n e^{ik_n x},$$

with

$$a_n = \left(\psi_n(x), \psi(x)\right) = \frac{1}{\sqrt{L}} \int_{-L/2}^{L/2} e^{-ik_n x} \psi(x)\,dx.$$

6.3 Infinite interval

Finally, we can consider the case where the functions are defined in an interval of infinite amplitude. In this case we can use the Fourier transform instead of the Fourier series. We define the Fourier transform for the function $\psi(x)$ as

$$A(k) = \frac{1}{\sqrt{2\pi}} \int_{-\infty}^{+\infty} \psi(x) e^{-ikx} dx,$$

while the inverse Fourier transform of the function $A(k)$ is

$$\psi(x) = \frac{1}{\sqrt{2\pi}} \int_{-\infty}^{+\infty} A(k) e^{ikx} dk.$$

Plancherel's theorem states that given a square summable function

$$\psi(x) \in \mathcal{L}^2(-\infty, +\infty),$$

there exists its Fourier transform (called $A(k)$ and shown above) which is a square summable function in the same interval and its inverse Fourier transform coincides (unless an almost everywhere null function) with the $\psi(x)$ as we have already written above equaling the inverse transform of $A(k)$ to the $\psi(x)$. We also remember the following representation of the Dirac delta distribution, which can be useful for some calculations with Fourier transforms,

$$\delta(x-x') = \frac{1}{2\pi} \int_{-\infty}^{+\infty} e^{ik(x-x')}.$$

The Parseval equation in this case adfirms that

$$\int_{-\infty}^{+\infty} |\psi(x)|^2 dx = \int_{-\infty}^{+\infty} |A(k)|^2 dk,$$

or, in equivalent form,

$$\Big(\psi(x), \psi(x)\Big) = \Big(A(x), A(x)\Big).$$

6.4 Coordinate and momentum space

We can say, putting together the results of the previous chapters, that the functions of the complete orthonormal system

$$\psi_n(x) = \frac{e^{ik_n x}}{\sqrt{L}}, \quad k_n = \frac{2\pi}{L} n \quad n \in \mathbb{Z},$$

coincide with the box normalized plane waves of the one-dimensional case and therefore represent free particles of defined momentum (and defined wave number k), confined in

6.4 Coordinate and momentum space

a spatial segment of amplitude L. Being confined they have quantized momentum as shown previously, i.e.

$$p_n = \hbar k_n = \frac{2\pi\hbar}{L} n, \quad n \in \mathbb{Z}.$$

Furthermore we recall that any square summable function in the interval $(-L/2, L/2)$ can be expanded in Fourier series, as a superposition of plane waves. Moreover, the square modulus of the Fourier coefficient

$$a_n = \left(\psi_n(x), \psi(x)\right) = \frac{1}{\sqrt{L}} \int_{-L/2}^{L/2} e^{-ik_n x} \psi(x)\, dx,$$

can be seen as the probability of finding the value of $\hbar k_n$ as the momentum of a particle described by the wave function $\psi(x)$.

Similarly the functions

$$\psi_k(x) = \frac{1}{\sqrt{2\pi}} e^{ikx}, \quad k \in \mathbb{R},$$

coincide with the plane waves of definite real k (monochromatic waves) defined in an unlimited (one-dimensional) space, without box normalization. So the previous expression (inverse transform of the ψ transform)

$$\psi_k(x) = \frac{1}{\sqrt{2\pi}} e^{ikx}, \quad k \in \mathbb{R},$$

can be seen as an overlap of these plane waves. In this case the role of the Fourier coefficient for the discrete case is played by the function $A(k)$ which is the Fourier transform of $\psi(x)$

$$\psi(x) = \frac{1}{\sqrt{2\pi}} \int_{-\infty}^{+\infty} A(k) e^{ikx} dk.$$

So its square modulus can be seen as the momentum probability density. Therefore we say that the quantity

$$|A(k)|^2 dk,$$

represents the probability of finding a momentum between

$$\hbar k$$

and

$$\hbar(k+dk),$$

6.4 Coordinate and momentum space

for a particle described by the wave function $\psi(x)$.

The obtained results can be trivially extended to the three-dimensional case. For example, the Fourier transforms would be

$$A(\vec{k}) = \frac{1}{(2\pi)^{3/2}} \int_{-\infty}^{+\infty} \psi(\vec{x}) e^{-i\vec{k}\cdot\vec{x}} d^3x$$

and

$$\psi(\vec{x}) = \frac{1}{(2\pi)^{3/2}} \int_{-\infty}^{+\infty} A(\vec{k}) e^{i\vec{k}\cdot\vec{x}} d^3k.$$

We now come to the concept of "coordinate space" and "momentum space". As seen so far we can say that the normalized wave function $\psi(x)$ represents a certain system in the coordinate space and in fact the quantity

$$|\psi(\vec{x})|^2 d^3x,$$

just gives the presence probability within the volume

$$(\vec{x}, \vec{x} + d^3x),$$

in a space with spatial coordinates (x, y, z). In a similar way we can say that the Fourier transform of $\psi(x)$, i.e. the function $A(k)$ (which is normalized to 1 as the ψ), represents the same system in the momentum space (sometimes called the wave vector space for obvious reasons). Indeed

$$|A(\vec{k})|^2 d^3k,$$

provides the probability of having a certain momentum within the "volume in the momentum space" given by

$$(\vec{p},\vec{p}+d^3p) = \hbar(\vec{k},\vec{k}+d^3k).$$

We can also write the Schrödinger equation in the momentum space for the function $A(k,t)$. This is given by

$$i\hbar\frac{\partial A(k,t)}{\partial t} = \frac{p^2}{2m}A(k,t) + V(i\hbar)\frac{\partial A(k,t)}{\partial p}.$$

The position and momentum operators in the coordinate space are

$$\hat{x} \to \vec{x}, \quad \hat{p} \to -i\hbar\vec{\nabla}$$

and, therefore, the first is a multiplicative operator, while the second is a differential one. The corresponding operators in

6.4 Coordinate and momentum space

the momentum space are

$$\hat{x} \to i\hbar \vec{\nabla}_{\vec{p}}, \quad \hat{p} \to \vec{p}$$

and, on the contrary, the first is a differential operator, while the second is a multiplicative one.

Chapter 7

Expectation value

We now introduce the concept of expectation value of a continuous observable (the discrete case provides for a summation rather than an integral). If we have a certain continuous observable A dependent on the variable x then, calling

$$P(a)\,da,$$

the probability of finding, by measuring A, a result between

$$(a, a+da),$$

then the expectation value of A is given by

$$\langle A \rangle = \int a P(a)\, da,$$

where the integral is extended to those values x on which we want to calculate the expectation value of A. We also define the root mean square as

$$\Delta A = \sqrt{\langle A^2 \rangle - \langle A \rangle^2},$$

which indicates how much the results of individual measurements deviate from the expectation value. The smaller this value, the closer the measurement results are to the more probable value. In case all the measurements always give the same value and the mean square deviation is exactly zero it is said that the quantity A is well determined, because measuring it several times always gives the same value. In quantum mechanics, where the presence probability density of a system is described by square modulus of the wave function, it is possible to calculate the expectation value of each function of the position, called $f(x)$, using

$$\langle f(\vec{x}) \rangle = \int f(\vec{x}) |\psi(\vec{x})|^2 d^3x.$$

In particular the expectation value of the position is

$$\langle \vec{x} \rangle = \int \vec{x} |\psi(\vec{x})|^2 d^3x,$$

while the expectation value of the square of the position is

$$\langle \vec{x}^2 \rangle = \int \vec{x}^2 |\psi(\vec{x})|^2 d^3x,$$

that can be written also as

$$\langle \vec{x}^2 \rangle = \int (x^2 + y^2 + z^2) |\psi(x,y,z)|^2 dxdydz.$$

Since that

$$|\psi(\vec{x})|^2 = \psi^*(\vec{x}) \psi(\vec{x}),$$

in general the expectation value of a function is also written as

$$\langle f(\vec{x}) \rangle = \int \psi^*(\vec{x}) f(\vec{x}) \psi(\vec{x}) d^3x,$$

which is the same as above, but it will be useful in this form when we need to calculate the expectation value of a non-multiplicative operator. The latter can be also written, in a compact way, using the scalar products notation

$$\langle f(\vec{x})\rangle = \Big(\psi(\vec{x}), f(\vec{x})\psi(\vec{x})\Big).$$

Chapter 8

Operators

In quantum mechanics the physical observables (such as the position, the momentum, the angular momentum, the energy) are represented by operators. Generally they are multiplicative or differential operators, depending on the space in which we want to make the calculations (coordinate space or momentum space). In most cases, where the coordinate space is chosen, we will see that the position operator is a multiplicative operator, while the momentum, angular momentum and kinetic energy operators are differential ones. Starting from here we will assume that we are working on the coordinate space, unless we explicitly state otherwise.

8. Operators

We said that a particle is ruled by its wave function (usually normalized to 1) and that all information on the particle can be obtained from it. The position probability density, for example, is given by the square modulus of the wave function, as mentioned so far. The importance of the operators lies in the fact that the application of an operator on a wave function (from the left) can return information on the physical quantity associated with the operator itself as we will see later.

8.1 Position operator

In the coordinate space the position operator is a multiplicative one. This operator can be written as

$$\hat{\vec{x}} = \vec{x},$$

or, in terms of its coordinates,

$$\hat{x}_i = x_i,$$

in agreement with the definition of its expectation value, i.e.

$$\langle \vec{x} \rangle = \int \Psi^*(\vec{x},t) \vec{x} \Psi(\vec{x},t) \, d^3x.$$

8.2 Momentum operator

To derive the momentum operator in the coordinate space we start from the definition of the expectation value of the wave vector $\vec{k}$ in the momentum space

$$\langle \vec{k} \rangle = \int \vec{k} |A(\vec{k},t)|^2 d^3k,$$

or also

$$\langle \vec{k} \rangle = \int A^*(\vec{k},t) \vec{k} A(\vec{k},t) \, d^3k.$$

We write the wave function in the momentum space $A(k,t)$ and its conjugate complex as a Fourier transform of the wave function in the coordinate space

$$A(\vec{k},t) = \frac{1}{(2\pi)^{3/2}} \int \Psi(\vec{x}',t) e^{i\vec{k}\cdot\vec{x}'} d^3x'$$

and

$$A^*(\vec{k},t) = \frac{1}{(2\pi)^{3/2}} \int \Psi^*(\vec{x}'',t) e^{-i\vec{k}\cdot\vec{x}''} d^3x''.$$

8. Operators

We replace these two quantities in the previous formula for the expectation value of k obtaining

$$\langle \vec{k} \rangle = \frac{1}{(2\pi)^3} \int \Psi^*(\vec{x}'',t) e^{i\vec{k}\cdot\vec{x}''} \cdot (i\vec{\nabla}_{\vec{x}'} e^{-i\vec{k}\cdot\vec{x}'}) \Psi(\vec{x}',t)$$
$$\cdot e^{-i\vec{k}\cdot\vec{x}'} d^3k \, d^3x' \, d^3x'',$$

having used

$$i\vec{\nabla}_{\vec{x}'} e^{-i\vec{k}\cdot\vec{x}'} = \vec{k} e^{-i\vec{k}\cdot\vec{x}'}.$$

We integrate by parts with respect to the variable x' so that

$$\langle \vec{k} \rangle = 0 - \frac{1}{(2\pi)^3} \int \Psi^*(\vec{x}'',t) e^{i\vec{k}\cdot\vec{x}''}$$
$$\cdot e^{-i\vec{k}\cdot\vec{x}'} [i\vec{\nabla}_{\vec{x}'} \Psi(\vec{x}',t)] d^3k \, d^3x' \, d^3x''.$$

We can now use one the following representation of the Dirac delta distribution

$$\delta^3(\vec{x}'' - \vec{x}') = \frac{1}{(2\pi)^3} \int e^{i\vec{k}\cdot(\vec{x}''-\vec{x}')} d^3k,$$

from which

$$\langle \vec{k} \rangle = \int \Psi^*(\vec{x}'',t) \delta^3(\vec{x}'' - \vec{x}') \cdot [-i\vec{\nabla}_{\vec{x}'} \Psi(\vec{x}',t)] d^3x' \, d^3x''.$$

8.2 Momentum operator

Calculating one of the two integrals, using the Dirac delta, and then changing the name of the remaining variable, calling it x for simplicity, we obtain

$$\langle \vec{k} \rangle = \int \Psi^*(\vec{x},t)[-i\vec{\nabla}_{\vec{x}}\Psi(\vec{x},t)]\, d^3x,$$

or

$$\langle \vec{k} \rangle = \int \Psi^*(\vec{x},t)(-i\vec{\nabla})\Psi(\vec{x},t)\, d^3x.$$

This implies, using the definition of expectation value in the coordinate space,

$$\langle \vec{k} \rangle = \int \Psi^*(\vec{x},t)\hat{\vec{k}}\Psi(\vec{x},t)\, d^3x,$$

that the operator to be associated with k in the coordinate space is

$$\hat{\vec{k}} = -i\vec{\nabla}, \quad \hat{k}_i = -i\frac{\partial}{\partial x_i}.$$

If we remember that

$$\vec{p} = \hbar \vec{k},$$

we obtain the operator

$$\hat{\vec{p}} = -i\hbar \vec{\nabla},$$

or
$$\hat{p}_i = -i\hbar \frac{\partial}{\partial x_i}$$

and therefore the expectation value can be written as

$$\langle \vec{k} \rangle = \int \Psi^*(\vec{x},t)(-i\hbar \vec{\nabla})\Psi(\vec{x},t)\,d^3x.$$

8.3 Energy operator

The energy, also called Hamiltonian, and indicated with the letter H, is given by the sum of the kinetic energy and the potential energy. The associated operator is

$$\hat{H} = \hat{T} + \hat{V},$$

where $\hat{T}$ is the kinetic energy operator and $\hat{V}$ is the potential energy operator. To obtain the expression of the former it is sufficient to observe that in non relativistic quantum mechanics the relation between the momentum p and the kinetic energy is

$$T = \frac{\vec{p}^2}{2m}$$

8.3 Energy operator

and therefore, knowing that the operator associated with the momentum is

$$\hat{\vec{p}} = -i\hbar \vec{\nabla},$$

we can deduce that

$$\hat{T} = \frac{\hat{\vec{p}}^2}{2m} = \frac{(-i\hbar\vec{\nabla})^2}{2m},$$

namely

$$\hat{T} = -\frac{\hbar^2 \vec{\nabla}^2}{2m},$$

which shows that the kinetic energy operator is a differential one. On the contrary the potential energy operator is a multiplicative one given by

$$\hat{V} = V(\vec{r}),$$

with

$$\vec{r} = \vec{x} = (x, y, z).$$

The energy operator is therefore given by

$$\hat{H} = -\frac{\hbar^2 \vec{\nabla}^2}{2m} + V(\vec{r}).$$

Notice how the stationary Schrödinger equation can be written as a differential equation in the following compact way (that is an eigenvalue equation for the energy operator $\hat{H}$)

$$\hat{H}\psi(\vec{r}) = E\psi(\vec{r}),$$

that is

$$\left(-\frac{\hbar^2\vec{\nabla}^2}{2m} + V(\vec{r})\right)\psi(\vec{r}) = E\psi(\vec{r}).$$

In the particular case of a free particle the potential energy is equal to zero $V = 0$ and one has to solve the stationary Schrödinger equation.

8.4 Angular momentum operator

The angular momentum is classically defined as the vector product

$$\vec{L} = \vec{r} \times \vec{p}.$$

Recalling now the operator expressions of the position vector $\vec{r}$ (multiplicative operator) and of the momentum vector $\vec{p}$

$$\hat{\vec{p}} = -i\hbar\vec{\nabla},$$

8.4 Angular momentum operator

we can write
$$\hat{\vec{L}} = \hat{\vec{r}} \times \hat{\vec{p}} = -i\hbar \vec{r} \times \vec{\nabla}.$$

Its components can be written using the Levi-Civita tensor (completely antisymmetric)

$$\varepsilon_{ijk} = \begin{cases} +1 & \text{per } ijk = 123 \text{ and cyclic permutations} \\ -1 & \text{per } ijk = 213 \text{ and cyclic permutations} \\ 0 & \text{if two or more indices are equal} \end{cases},$$

where, for example, 231 and 312 are cyclic permutations of 123. Using this tensor we can write the generic component of the vector product between any two vectors as

$$(\vec{a} \times \vec{b})_i = \varepsilon_{ijk} a_j b_k$$

where a summation over repeated indices is understood (Einstein notation) and therefore for the angular momentum

$$\hat{L}_i = \varepsilon_{ijk} x_j \hat{p}_k = -i\hbar \varepsilon_{ijk} x_j \frac{\partial}{\partial x_k}.$$

We explicitly write the three Cartesian components of the angular momentum operator

$$\hat{L}_1 = \hat{L}_x = -i\hbar \left(y\frac{\partial}{\partial z} - z\frac{\partial}{\partial y} \right),$$

$$\hat{L}_2 = \hat{L}_y = -i\hbar \left(z\frac{\partial}{\partial x} - x\frac{\partial}{\partial z} \right)$$

and

$$\hat{L}_3 = \hat{L}_z = -i\hbar \left(x\frac{\partial}{\partial y} - y\frac{\partial}{\partial x} \right).$$

The square of the angular momentum is another important operator used in quantum mechanics and it is trivially given by

$$\hat{L}^2 = \hat{\vec{L}} \cdot \hat{\vec{L}} = \hat{L}_x^2 + \hat{L}_y^2 + \hat{L}_z^2.$$

8.5 Spherical coordinates

The operators introduced so far can also be written in spherical coordinates. We can define them as

$$\begin{cases} x = r\sin\theta\cos\phi \\ y = r\sin\theta\sin\phi \\ z = r\cos\theta \end{cases},$$

8.5 Spherical coordinates

with
$$r \geq 0, \quad 0 \leq \theta \leq \pi, \quad 0 \leq \phi \leq 2\pi,$$

where we remember that
$$r^2 = x^2 + y^2 + z^2.$$

The choice of spherical coordinates is particularly useful for central problems, as we will see later. We want to obtain the expressions in spherical coordinates of the operators introduced in the last chapter. For instance we observe that the partial derivatives with respect to the Cartesian coordinates can be written as

$$\frac{\partial}{\partial x} = \frac{\partial r}{\partial x}\frac{\partial}{\partial r} + \frac{\partial \theta}{\partial x}\frac{\partial}{\partial \theta} + \frac{\partial \phi}{\partial x}\frac{\partial}{\partial \phi},$$

$$\frac{\partial}{\partial y} = \frac{\partial r}{\partial y}\frac{\partial}{\partial r} + \frac{\partial \theta}{\partial y}\frac{\partial}{\partial \theta} + \frac{\partial \phi}{\partial y}\frac{\partial}{\partial \phi}$$

and
$$\frac{\partial}{\partial z} = \frac{\partial r}{\partial z}\frac{\partial}{\partial r} + \frac{\partial \theta}{\partial z}\frac{\partial}{\partial \theta} + \frac{\partial \phi}{\partial z}\frac{\partial}{\partial \phi},$$

The partial derivatives of the spherical coordinates with respect to the Cartesian ones are easily obtained from the transformations between the spherical coordinates and the Cartesian ones, shown above. The third component of the angular momentum operator is

$$\hat{L}_3 = \hat{L}_z = -i\hbar \left(x \frac{\partial}{\partial y} - y \frac{\partial}{\partial x} \right).$$

After some algebraic simplifications, we arrive at

$$\hat{L}_z = -i\hbar \frac{\partial}{\partial \phi}.$$

For the choice made when we wrote the transformation between spherical and Cartesian coordinates (i.e. having chosen the z axis as the polar axis) the expression of the component along z of the angular momentum (infinitesimal rotation around its axis) is the simplest one. In fact the other two components of the angular momentum are

$$\hat{L}_x = -i\hbar \left(\sin\phi \frac{\partial}{\partial \theta} + \frac{\cos\phi}{\tan\theta} \frac{\partial}{\partial \phi} \right)$$

8.5 Spherical coordinates

and
$$\hat{L}_y = -i\hbar \left(-\cos\phi \frac{\partial}{\partial \theta} + \frac{\sin\theta}{\tan\theta} \frac{\partial}{\partial \phi} \right).$$

The square of the angular momentum, using the obtained results, is represented by the following operator in spherical coordinates

$$\hat{L}^2 = l\hbar \left(\frac{1}{\tan\theta} \frac{\partial}{\partial \theta} + \frac{\partial^2}{\partial \theta^2} + \frac{1}{\sin^2\theta} \frac{\partial^2}{\partial \phi^2} \right).$$

For central problems we can write the square of the momentum operator as

$$\hat{p}^2 = \hat{p}_r^2 + \frac{\hat{L}^2}{r^2} = \left(\frac{1}{2} \left(\frac{\vec{r} \cdot \hat{\vec{p}}}{r} + \frac{\hat{\vec{p}} \cdot \vec{r}}{r} \right) \right)^2 + \frac{\hat{L}^2}{r^2},$$

where the radial component of the momentum has been written as an average between the two possible ways of classically writing it. In fact, the operators of r and p (unlike the corresponding classical vector quantities) do not commute, in the sense that the order in which they are written in the product is important. Furthermore this the only way in which the radial component of the momentum is correctly an Hermitian operator. In fact in quantum mechanics all operators of observables are Hermitians in order to obtain real eigenvalues that represent the physical quantities. Explicitly, remembering the expression

$$\hat{\vec{p}} = -i\hbar \vec{\nabla},$$

we can calculate

$$\hat{p}_r = -i\hbar \left(\frac{1}{r} + \frac{\partial}{\partial r} \right)$$

and

$$\hat{p}_r^2 = -\hbar^2 \left(\frac{2}{r} \frac{\partial}{\partial r} + \frac{\partial^2}{\partial r^2} \right).$$

Finally, using the previous expression of p, we obtain

$$\hat{p}^2 = -\hbar^2 \nabla^2 = -\hbar^2 \Delta,$$

with the Laplace operator expressed in spherical coordinates. The kinetic energy operator has the form

$$\hat{T} = -\frac{\hbar^2 \Delta}{2m}.$$

Chapter 9

Commutation relations

We give the definition of the commutator between two operators. Given two operators A and B their commutator is

$$[\hat{A}, \hat{B}] = \hat{A}\hat{B} - \hat{B}\hat{A}.$$

Similarly their anticommutator is defined as

$$\{\hat{A}, \hat{B}\} = \hat{A}\hat{B} + \hat{B}\hat{A}.$$

If the commutator between two operators is null then they commute. We now explicitly calculate the commutator between some common operators. First of all, we observe trivially that the position operator commute with itself (each component commute with any other) and the same apply for the components of the momentum operator. Indeed

$$[\hat{x}_i, \hat{x}_j] = \hat{x}_i \hat{x}_j - \hat{x}_j \hat{x}_i = 0$$

and

$$[\hat{p}_i, \hat{p}_j] = -\hbar^2 \left(\frac{\partial^2}{\partial_i \partial_j} - \frac{\partial^2}{\partial_j \partial_i} \right) = 0.$$

On the other hand if we calculate the commutator between a component of the position operator and a component of the momentum operator we get (when calculating a commutator it is always necessary to think as it is applied to a generic function)

$$\begin{aligned}[] [\hat{x}_i, \hat{p}_j] \psi &= \left[x_i, -i\hbar \frac{\partial}{\partial x_j} \right] \psi = -i\hbar x_i \frac{\partial \psi}{\partial x_j} \\ &+ i\hbar \frac{\partial (x_i \psi)}{\partial x_j} = i\hbar \frac{\partial x_i}{\partial x_j} \psi, \end{aligned}$$

from which the operator identity

$$[\hat{x}_i, \hat{p}_j] = i\hbar \delta_{ij}.$$

So for $i = j$, i.e. for canonically conjugated variables, the commutator is $i\hbar$. In general, using the commutators property

$$[\hat{A}, \hat{B}\hat{C}] = [\hat{A}, \hat{B}]\hat{C} + \hat{B}[\hat{A}, \hat{C}],$$

for a pair of canonically conjugated variables (which we simply call x and p) we have

$$[\hat{x}^n, \hat{p}] = i\hbar n \hat{x}^{n-1}, \quad n \in \mathbb{N}^+$$

and

$$[\hat{x}, \hat{p}^n] = i\hbar n \hat{p}^{n-1}, \quad n \in \mathbb{N}^+.$$

Furthermore, given a function that can be expanded in series of powers

$$f(x) = \sum_n a_n x^n,$$

we have

$$[f(\hat{x}), \hat{p}] = i\hbar \frac{df(\hat{x})}{d\hat{x}},$$

as seen for example from

$$[f(\hat{x}),\hat{p}] = \sum_n a_n[\hat{x}^n,\hat{p}] = i\hbar \sum_n a_n n \hat{x}^{n-1}$$

and similarly

$$[\hat{x}, f(\hat{p})] = i\hbar \frac{df(\hat{p})}{d\hat{p}}.$$

Using all these results, we can calculate the commutator between the potential operator and the momentum one as

$$[\hat{V}(\vec{x}), \hat{p}_i] = i\hbar \frac{\partial \hat{V}(\vec{x})}{\partial x_i}$$

and the commutator between the position operator and the kinetic energy one

$$[\hat{x}, \hat{T}] = \frac{[\hat{x}_i, \sum_j \hat{p}_j^2]}{2m} = \frac{[\hat{x}_i, \hat{p}_i^2]}{2m} = \frac{i\hbar \hat{p}_i}{m}.$$

We also remember that obviously

$$[\hat{A}, \hat{B}] = -[\hat{B}, \hat{A}].$$

We can say that the components of x do not commute with the Hamiltonian H, while those of p commute with H only for particular potential that are constants in some variable x. With calculations analogous to those made to derive the commutator between x and p we can calculate the commutator between any two components of the angular momentum as

$$[\hat{L}_i, \hat{L}_j] = i\hbar \varepsilon_{ijk} L_k,$$

from which, for example,

$$[\hat{L}_x, \hat{L}_y] = i\hbar L_z.$$

We observe that

$$\begin{aligned}[\hat{L}_i, \hat{L}^2] &= [\hat{L}_i, \hat{L}_j \hat{L}_j] = \hat{L}_j [\hat{L}_i, \hat{L}_j] + [\hat{L}_i, \hat{L}_j] \hat{L}_j \\ &= i\hbar \varepsilon_{ijk} \{\hat{L}_j, \hat{L}_k\} = 0,\end{aligned}$$

in fact in the final product there is a summation on the indices j and k where the first factor (the Levi-Civita tensor) is antisymmetric for the exchange of j and k, while the second factor (the anticommutator) is symmetric for the exchange of j and k. So we can write

$$[\hat{L}_i, \hat{L}^2] = 0.$$

Moreover, with similar calculations, we obtain

$$[\hat{L}_i, \hat{x}_j] = i\hbar \varepsilon_{ijk} \hat{x}_k$$

and

$$[\hat{L}_i, \hat{p}_j] = i\hbar \varepsilon_{ijk} \hat{p}_k \,,$$

Finally

$$[\hat{L}_i, \hat{r}^2] = 0\,, \qquad [\hat{L}_i, \hat{p}^2] = 0\,.$$

Chapter 10

Uncertainty principle

We begin this chapter considering the simultaneous measure of two observables. Two Hermitian operators A and B admit a common set of eigenfunctions if and only if they commute, that is, if and only if their commutator is null

$$[\hat{A}, \hat{B}] = 0.$$

This means that two observables can be known simultaneously if the associated operators commute. The uncertainty principle is a theorem related to the simultaneous measure of two observables. We define the uncertainty of a measure (on a generic state described by the wave function ψ) the

following
$$\Delta A = \sqrt{\langle A^2 \rangle - \langle A \rangle^2}.$$

To obtain the statement of the uncertainty principle we consider two observables A and B represented by Hermitian operators. The commutator between A and B is antihermitian and so the quantity
$$i[\hat{A}, \hat{B}],$$
will be Hermitian. Therefore its expected value on a generic state described by the wave function ψ is real and it is expressed as
$$\langle i[\hat{A}, \hat{B}] \rangle_\psi \in \mathbb{R}.$$

We can write the following relation
$$\begin{aligned}\langle i[\hat{A}, \hat{B}] \rangle^2_\psi &= |\langle i[\hat{A}, \hat{B}] \rangle_\psi|^2 = |\langle AB \rangle_\psi - \langle BA \rangle_\psi|^2 \\ &\leq (|\langle AB \rangle_\psi| + |\langle BA \rangle_\psi|)^2 = (2|(\psi, AB\psi)|)^2,\end{aligned}$$

in fact, being A and B Hermitian,
$$\begin{aligned}|\langle BA \rangle_\psi| &= |(\psi, BA\psi)| = |(\psi, BA\psi)^*| \\ &= |(BA\psi, \psi)| = |(\psi, AB\psi)|.\end{aligned}$$

Always using the fact that A and B are Hermitian the previous inequality becomes

$$|\langle i[A,B]\rangle_\psi|^2 \leq 4|(A\psi,B\psi)|^2 \leq 4(\psi,A^2\psi)(\psi,B^2\psi),$$

where Schwarz's inequality was also used. This result can also be written as

$$|\langle i[A,B]\rangle_\psi|^2 \leq 4\langle A^2\rangle_\psi \langle B^2\rangle_\psi$$

and must apply to every pair of Hermitian operators. We make the replacements with the two new Hermitian operators

$$A \to A - \langle A\rangle_\psi$$

and

$$B \to B - \langle B\rangle_\psi.$$

In this case, being

$$[A - \langle A\rangle_\psi, B - \langle B\rangle_\psi] = [A,B],$$

we obtain

$$|\langle i[A,B]\rangle_\psi|^2 \leq 4\langle (A-\langle A\rangle_\psi)^2\rangle_\psi \langle (B-\langle B\rangle_\psi)^2\rangle_\psi.$$

We observe that

$$\begin{aligned}\langle (A-\langle A\rangle_\psi)^2\rangle_\psi &= \langle A^2\rangle_\psi + \langle A\rangle_\psi^2 - 2\langle A\rangle_\psi^2 \\ &= \langle A^2\rangle_\psi - \langle A\rangle_\psi^2 = (\Delta A)^2.\end{aligned}$$

Therefore

$$|\langle i[A,B]\rangle_\psi|^2 \leq 4(\Delta A)^2(\Delta B)^2,$$

from which

$$|\langle i[A,B]\rangle_\psi|^2 \leq 2\Delta A\,\Delta B.$$

The uncertainty principle for two Hermitian operators A and B can be finally written as

$$\Delta A\,\Delta B \geq \frac{1}{2}|\langle i[A,B]\rangle_\psi|^2.$$

A particular example is that of two canonically conjugated operators Q and P. By placing $A=Q$ and $B=P$, being

$$[Q,P]=i\hbar,$$

we obtain the known inequality

$$\Delta Q \Delta P \geq \frac{\hbar}{2}.$$

Finally, we observe that the generic inequality

$$\Delta A \, \Delta B \geq \frac{1}{2} |\langle i[A,B]\rangle_\psi|^2,$$

allows to have (for a particular state)

$$\Delta A = 0, \quad \Delta B = 0,$$

even when the two operators do not commute, if

$$\langle [A,B]\rangle_\psi = 0.$$

There is also an uncertainty relation between energy and time, i.e.

$$\Delta E \, \Delta t \geq \frac{\hbar}{2}.$$

Chapter 11

Eigenvalue equations

If we perform a measurement on a physical system described by a wave function we obtain a real number representative of the measured quantity. In quantum mechanics this number is obtained by applying the operator associated with the observable (which you want to measure) on the wave function that describes the system. Given an operator A we can write the eigenvalue equation

$$\hat{A}\psi_i = a_i\psi_i,$$

where the eigenfunctions of the operator A (eigenvectors) must be functions that describe the state of a physical system. The set of eigenvalues of the operator A is called the spectrum of A and can be discrete or continuous. If an eigenvalue corresponds to a single eigenfunction then it is said to be non-degenerate, otherwise it is said to be degenerate of order n, where n is the number of linearly independent eigenfunctions that have that same eigenvalue. To summarize, we say that an eigenfunction always corresponds to a single eigenvalue, while an eigenvalue can have multiple associated eigenfunctions (if degenerate). In quantum mechanics usually we work with Hermitian operators, because they have particular properties, such as

- the eigenvalues of Hermitian operators are real;
- the eigenfunctions of a Hermitian operator belonging to distinct eigenvalues are orthogonal.

11.1 Position operator

The eigenvalue equation for the position operator (we consider a generic component of index k) can be written as (the tilde indicates the eigenvalue)

$$\hat{x}\psi_{\tilde{x}_k}(x_k) = \tilde{x}\psi_{\tilde{x}_k}(x_k)$$

11.1 Position operator

and, in this case, the only eigenfunction that satisfies this equation is the generalized Dirac delta function that "selects" the position of the particle. This eigenfunction is written, unless a multiplicative constant, as

$$\psi_{\tilde{x}_k}(x_k) = \delta(x_k - \tilde{x}_k),$$

or, in three dimensions,

$$\psi_{\tilde{\vec{x}}}(\vec{x}) = \delta^3(\vec{x} - \tilde{\vec{x}}).$$

We observe that the spectrum is continuous (there are no limitations on the real values for the eigenvalues) and not degenerate (for each eigenvalue there is only one eigenfunction). This eigenfunction can be normalized in a broad sense and the normalization, for operators with a continuous spectrum, is given by placing

$$\left(\psi_{\tilde{x}_k}, \psi_{\tilde{x}'_k}\right) = \delta(\tilde{x}_k - \tilde{x}'_k).$$

The set of eigenfunctions forms a complete orthonormal set in a broad sense.

11.2 Momentum operator

The eigenvalue equation for the x component of the momentum operator in the coordinate space is the differential equation

$$\hat{p}_x \psi_{p_x}(x) = p_x \psi_{p_x}(x),$$

namely

$$-i\hbar \frac{\partial}{\partial x} \psi_{p_x}(x) = p_x \psi_{p_x}(x).$$

Its solution leads to

$$\frac{d\psi_{p_x}(x)}{\psi_{p_x}(x)} = \frac{i}{\hbar} p_x dx,$$

from which finally

$$\psi_{p_x}(x) = N e^{i p_x x/\hbar},$$

where N is an arbitrary constant that can be used for the normalization. The wave function for a free particle with fixed component x of the momentum is a plane wave, as discussed also previously. The spectrum is continuous and not degenerate. The constant N can be fixed by imposing a normalization

11.2 Momentum operator

in the broad sense as for the position operator. For example we can impose that

$$\left(\psi_{p_x}, \psi_{p'_x}\right) = \delta(p_x - p'_x).$$

In this case, by calculating, we have

$$\left(\psi_{p_x}, \psi_{p'_x}\right) = 2\pi\hbar|N|^2\delta(p_x - p'_x),$$

from which it follows that we can choose N real and equal to

$$N = \frac{1}{\sqrt{2\pi\hbar}}.$$

The eigenfunctions become

$$\psi_{p_x}(x) = \frac{1}{\sqrt{2\pi\hbar}} e^{ip_x x/\hbar}$$

and, in three dimensions,

$$\psi_{\vec{p}}(\vec{x}) = \frac{1}{(2\pi\hbar)^{3/2}} e^{i\vec{p}\cdot\vec{x}/\hbar}.$$

These eigenfunctions represent a complete orthonormal system, in a broad sense, for the class of square summable function. For the x component of the wave vector operator, similarly to what has been done previously, we obtain the eigenvalue equation

$$\hat{k}_x \psi_{k_x}(x) = -i\hbar \frac{\partial}{\partial x} \psi_{k_x}(x) = k_x \psi_{k_x}(x),$$

from which

$$\psi_{k_x}(x) = \frac{1}{\sqrt{2\pi}} e^{ik_x x}$$

and, in three dimensions,

$$\psi_{\vec{k}}(\vec{x}) = \frac{1}{(2\pi)^{3/2}} e^{i\vec{k} \cdot \vec{x}}.$$

11.3 The operator $\hat{L}_z$

In spherical coordinates, the third component of the angular momentum operator is

$$\hat{L}_z = -i\hbar \frac{\partial}{\partial \phi}.$$

This operator acts only on the functions that depend on the angle ϕ and therefore for its eigenfunctions we can write the

11.3 The operator $\hat{L}_z$

relation
$$\psi(\phi) = \psi(\phi + 2\pi).$$

The eigenvalue equation for the third component of the angular momentum operator is written as

$$\hat{L}_z \psi(\phi) = -i\hbar \frac{\partial \psi(\phi)}{\partial \phi} = l_z \psi(\phi),$$

where l_z are the eigenvalues. Solving it by variable separation we get

$$\frac{d\psi}{\psi} = \frac{il_z}{\hbar} d\phi$$

and

$$\psi(\phi) = C e^{il_z \phi/\hbar},$$

where C is the integration constant. Applying the above periodicity condition we obtain

$$C e^{il_z \phi/\hbar} = C e^{il_z(\phi + 2\pi)/\hbar}$$

and therefore

$$1 = e^{2\pi i l_z/\hbar},$$

from which
$$\frac{2\pi l_z}{\hbar} = 2\pi m, \quad m \in \mathbb{Z},$$

Simplifying we obtain the quantized eigenvalues

$$l_z = m\hbar, \quad m \in \mathbb{Z}.$$

The eigenfunctions are

$$\psi_m(\phi) = C e^{im\phi}$$

and we observe that the spectrum is non-degenerate, with an eigenfunction for each eigenvalue. To find the constant C with the normalization we calculate the scalar product

$$(\psi_m, \psi_m) = |C|^2 \int_0^{2\pi} d\phi \, 1 = 2\pi |C|^2,$$

hence
$$C = \frac{1}{\sqrt{2\pi}}.$$

In addition, the eigenfunctions are orthogonal being

$$(\psi_m, \psi_{m'}) = |C|^2 \int_0^{2\pi} d\phi \, e^{i(m-m')\phi} = |C|^2 2\pi \delta_{mm'}.$$

11.3 The operator $\hat{L}_z$

The normalized eigenfunctions of the third component of the angular momentum operator are, finally,

$$\psi_m(\phi) = \frac{1}{\sqrt{2\pi}} e^{im\phi},$$

where we remember that $m \in \mathbb{Z}$. These eigenfunctions form a complete orthonormal system (for the square summable functions).

Part II

Particle Physics

Chapter 12

Introduction

This second part is an introduction to particle physics. The main topics are:
- the special relativity;
- the relativistic kinematic;
- the elementary particles (quark and leptons);
- the quark model;
- the hadrons;
- the cosmic rays;
- the energy loss;
- the strangeness;
- the parity;

- the charge conjugation;
- the baryon and lepton numbers;
- the isospin;
- the hypercharge,
- the *G*-parity;
- the helicity;
- the chirality;
- scattering and diffusion;
- the cross section;
- the particle decays.

Chapter 13

Natural units

We use the natural units where

$$\hbar = c = 1.$$

So for example

$$\hbar c = 1, \qquad h = 2\pi.$$

This simplifies many formulas, for example

$$E_0 = mc^2 \rightarrow E_0 = m.$$

Energy and mass are both measured in electronvolt (eV) and its multiples. It is also useful to remember that

$$\hbar c \approx 197 \text{ MeV fm}$$

and therefore in natural units

$$1\, fm^{-1} \approx 197 \text{ MeV},$$

or

$$1\, MeV^{-1} \approx 197 \text{ fm}.$$

In summary we have the following units

$$[L] = [t] = \text{eV}^{-1},$$

$$[m] = [E] = \text{eV}.$$

Chapter 14

Bases of relativity

14.1 Four-vectors

A four-vector can be written, in a four-dimensional space, by means of its contravariant components

$$a^\mu = (a^0, a^1, a^2, a^3).$$

The metric tensor, in a Minkowsky space, can be written a

$$\eta_{\mu\nu} = \text{diag}(+1,-1,-1,-1)$$
$$= \begin{pmatrix} +1 & 0 & 0 & 0 \\ 0 & -1 & 0 & 0 \\ 0 & 0 & -1 & 0 \\ 0 & 0 & 0 & -1 \end{pmatrix},$$

hence the scalar product between two four-vector a and b

$$\begin{aligned} a \cdot b &= \sum_{\mu,\nu=0}^{3} \eta_{\mu\nu} a^{\mu} b^{\nu} = \eta_{\mu\nu} a^{\mu} b^{\nu} = a_{\nu} b^{\nu} \\ &= a^0 b^0 - a^1 b^1 - a^2 b^2 - a^3 b^3 \\ &= a^0 b^0 - \vec{a} \cdot \vec{b}, \end{aligned}$$

where we have used the Einstein notation for repeated indices and where the vectors $\vec{a}$ and $\vec{b}$ refer only to the spatial component of a and b, i.e.

$$\vec{a} = (a^1, a^2, a^3)$$

and

$$\vec{b} = (b^1, b^2, b^3).$$

14.2 Lorentz transformations

Consider two four-dimensional reference frames. In the first one let be an event represented by the four-vector

$$x^\mu = \left(x^0, x^1, x^2, x^3\right).$$

We denote the four-vector for the same event on the second frame with

$$x'^\mu = \left(x'^0, x'^1, x'^2, x'^3\right).$$

A linear and homogeneous transformation between the two frames can be written as

$$x'^\mu = \Lambda^\mu_\nu x^\nu. \qquad (14.2.1)$$

where the Λ^μ_ν are the components of the matrix

$$\Lambda = \begin{pmatrix} \Lambda^0_0 & \Lambda^0_1 & \Lambda^0_2 & \Lambda^0_3 \\ \Lambda^1_0 & \Lambda^1_1 & \Lambda^1_2 & \Lambda^1_3 \\ \Lambda^2_0 & \Lambda^2_1 & \Lambda^2_2 & \Lambda^2_3 \\ \Lambda^3_0 & \Lambda^3_1 & \Lambda^3_2 & \Lambda^3_3 \end{pmatrix}.$$

If this transformation leaves the interval unchanged

$$ds^2 = dx_\sigma dx^\sigma = \eta_{\rho\sigma} dx^\rho dx^\sigma,$$

then it is said homogeneous Lorentz transformation. So you have

$$ds'^2 = \eta_{\mu\nu} dx'^\mu dx'^\nu = ds^2 = \eta_{\rho\sigma} dx^\rho dx^\sigma$$

and, using the expressions

$$dx'^\mu = \Lambda^\mu_\rho dx^\rho,$$
$$dx'^\nu = \Lambda^\nu_\sigma dx^\sigma,$$

we obtain

$$\eta_{\mu\nu} \Lambda^\mu_\rho \Lambda^\nu_\sigma dx^\rho dx^\sigma = \eta_{\rho\sigma} dx^\rho dx^\sigma,$$

from which

$$\eta_{\mu\nu} \Lambda^\mu_\rho \Lambda^\nu_\sigma = \eta_{\rho\sigma}.$$

This is the condition for the Λ matrices in order to represent a homogeneous Lorentz transformation. Finally we observe

that, being,
$$dx'^\mu = \frac{\partial x'^\mu}{\partial x'^\nu} dx^\nu,$$

we can write, using the Eq. (14.2.1) for dx^μ,
$$\frac{\partial x'^\mu}{\partial x'^\nu} dx^\nu = \Lambda^\mu_\nu dx^\nu,$$

or
$$\Lambda^\mu_\nu = \frac{\partial x'^\mu}{\partial x'^\nu}.$$

14.3 Relativistic kinematics

The energy-momentum four-vector has components
$$p^\mu = (E/c, \vec{p}) = (m\gamma c, m\gamma \vec{v})$$

and
$$p_\mu = \eta_{\mu\nu} p^\nu = (E/c, -\vec{p}) = (m\gamma c, -m\gamma \vec{v}),$$

where E is the energy, m is the mass, $\vec{p}$ is the momentum vector and γ is the Lorentz factor, given by
$$\gamma = \frac{1}{\sqrt{1 - v^2/c^2}}.$$

We can calculate the Lorentz invariant quantity

$$p^2 = p_\mu p^\mu = m^2 \gamma^2 c^2 - m^2 \gamma^2 \vec{v}^2 = m^2 c^2,$$

moreover

$$p^2 = \frac{E^2}{c^2} - \vec{p}^2$$

therefore we have the relation, called mass-shell, valid for a free particle

$$E^2 = m^2 c^4 + \vec{p}^2 c^2.$$

For zero mass particles, such as the photon,

$$E = pc,$$

with $p = |\vec{p}|$. We also remember that

$$\vec{\beta} = \vec{v}/c.$$

Using the previous results, the following useful relationships can be obtained, in natural units and valid for free particles,

$$\begin{cases} E^2 = p^2 + m^2 \\ E = \gamma m = T + m \\ p = \beta E \\ T = (\gamma - 1)m \end{cases},$$

where T is the kinetic energy. Combining these we have also

$$p = \gamma \beta m.$$

14.4 Invariant mass

The mass of a system of N non-interacting particles, each of energy E_i and momentum $\vec{p}_i$ is called the invariant mass and it is written as

$$m = \sqrt{\left(\sum_{i=1}^{N} E_i\right)^2 - \left(\sum_{i=1}^{N} \vec{p}_i\right)^2}.$$

Chapter 15

Particles

In general, particles can be divided, depending on whether their spin (in units of $\hbar$) is integer or semi-integer, in bosons and fermions. Examples of particles with semi-integer spin, called fermions, are the quarks, the leptons, the proton and the neutron. Examples of bosons are the photon and the gluon. Fermions obey the so-called Fermi-Dirac statistic, while bosons obey the Bose-Einstein statistic. In particular, fermions obey the Pauli exclusion principle, due to the antisymmetric wave function of a system of identical fermions.

15.1 Elementary particles

The standard model of the elementary particles includes six leptons, six antileptons, six quarks and six antiquarks divided into three families in addition to the gauge bosons mediating the interactions.

15.1.1 Quarks

There are twelve quarks: up, down, charm, strange, top, bottom and their antiparticles. They can be divided into doublets

$$\begin{pmatrix} u \\ d \end{pmatrix}, \quad \begin{pmatrix} c \\ s \end{pmatrix}, \quad \begin{pmatrix} t \\ b \end{pmatrix},$$

similarly, for the antiparticles,

$$\begin{pmatrix} \bar{u} \\ \bar{d} \end{pmatrix}, \quad \begin{pmatrix} \bar{c} \\ \bar{s} \end{pmatrix}, \quad \begin{pmatrix} \bar{t} \\ \bar{b} \end{pmatrix},$$

The quark properties are summarized in Table 15.1.1.

15.1.2 Leptons

There are twelve leptons: electron, electronic neutrino, muon, muonic neutrino, tauone, tauonic neutrino and their antipar-

15.1 Elementary particles

Table 15.1.1: *Charge and mass of quarks.*

quark	Q (e)	m (MeV)
up (u)	+2/3	1.7 − 3.3
down (d)	−1/3	4.1 − 5.8
charm (c)	+2/3	1180 − 1340
strange (s)	−1/3	80 − 130
top (t)	+2/3	173100 ± 1300
bottom (b)	−1/3	4130 − 4370

ticles. They can be divided into doublets

$$\begin{pmatrix} e^- \\ \nu_e \end{pmatrix}, \quad \begin{pmatrix} \mu^- \\ \nu_\mu \end{pmatrix}, \quad \begin{pmatrix} \tau^- \\ \nu_\tau \end{pmatrix},$$

similarly, for the antiparticles,

$$\begin{pmatrix} e^+ \\ \overline{\nu}_e \end{pmatrix}, \quad \begin{pmatrix} \mu^+ \\ \overline{\nu}_\mu \end{pmatrix}, \quad \begin{pmatrix} \tau^+ \\ \overline{\nu}_\tau \end{pmatrix},$$

The properties of leptons are summarized in Table 15.1.2.

15.1.3 Quark model

The quark model assumes that mesons and baryons are composed by valence quarks. A quark-antiquark pair for a meson

Table 15.1.2: *Charge and mass of leptons.*

lepton	Q (e)	m (MeV)
e^-	-1	0.511
ν_e	0	$< 0.22 \cdot 10^{-6}$
μ^-	-1	105.66
ν_μ	0	< 0.17
τ^-	-1	1777
ν_τ	0	< 15.5

and three quarks (or three antiquarks) for a baryon. The symmetry of isospin is based on the $SU(2)$ group and the quark model extends it by considering the $SU(3)$ group. The model is based on the three light quarks: up, down and strange. Quarks are characterized by fractional charge and baryon number and cannot be observed isolated. There are two fundamental representations which are denoted by **3** for the three quarks and by $\overline{\mathbf{3}}$ for the respective antiquarks. The two representations are schematized in Figure 15.1.1 and 15.1.2.

15.2 Fundamental interactions

In nature there are four elementary forces mediated by the corresponding gauge bosons. These forces are, in order of intensity:

15.2 Fundamental interactions

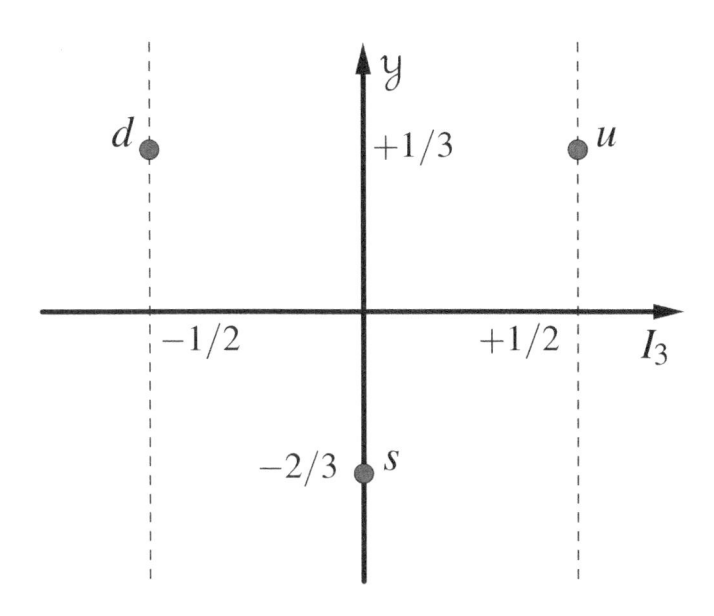

Figure 15.1.1: *Rappresentazione 3 per i quark u, d, s.*

- **Strong nuclear force**: holds together quarks to form composite particles, the hadrons. The mediator is the gluon g which has spin and parity $J^P = 1^-$;
- **Electromagnetic force**: acts on electrically charged particles. The mediator is the photon γ which has spin and parity $J^P = 1^-$;
- **Weak nuclear force**: appears mainly in radioactive decays and it is responsible for flavour quark change. The mediators are the $W^\pm$ with spin and parity $J^P = 1^-$ and the Z^0

Figure 15.1.2: *Rappresentazione $\bar{3}$ per i quark $\bar{u}, \bar{d}, \bar{s}$.*

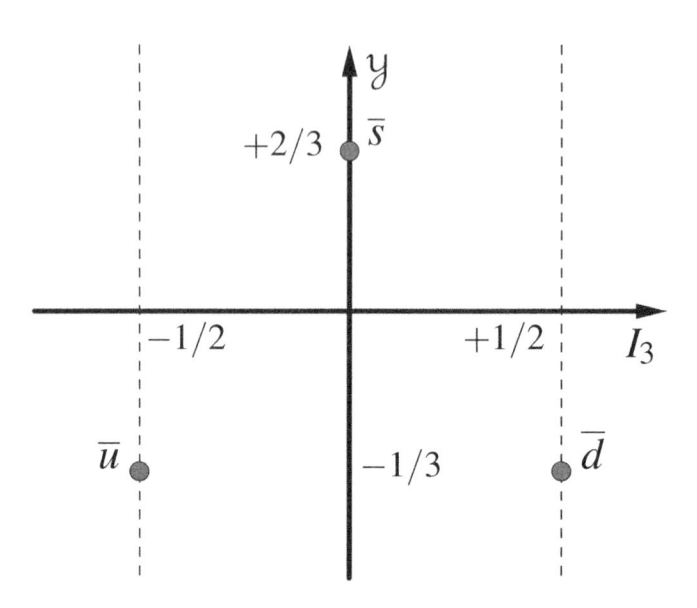

with spin and parity $J^P = 1^+$;

- **Gravitational force**: acts on particles with mass. In particle physics it is often negligible. Its mediator has not yet been experimentally identified, it is said graviton and would have spin and parity $J^P = 2^+$.

With reference to the first three forces, most important in particle physics, we have the behavior of leptons and quarks shown in Table 15.2.1.

15.3 Hadrons

Table 15.2.1: *Interactions and elementary particles*

particle	strong int.	EM int.	weak int.
quark	YES	YES	YES
charge leptons	NO	YES	YES
neutral leptons	NO	NO	YES

15.3 Hadrons

In general, hadrons are particles composed of quarks (q) or antiquarks ($\bar{q}$). They are mainly divided into mesons ($q\bar{q}$) and baryons (qqq or $\overline{qqq}$).

15.3.1 Mesons

Mesons are non-elementary particles and their name originates by their mass that has values between the electron mass and the proton mass. They generally decay into photons or leptons. They have integer spin and are therefore bosons. In the quark model they are formed by a quark-antiquark pair, $q\bar{q}$ and they belong to the multiplets that originate from the product of the representations

$$3 \otimes \bar{3} = 1 \oplus 8,$$

i.e. they belong to an octet and to a singlet, a nonet overall. In the ground state ($L = 0$), there are mesons with $J^P = 0^-$ or $J^P = 1^-$. The mesons with $J^P = 0^-$ can be represented in the nonet shown in Figure 15.3.1, while those with $J^P = 1^-$ can be represented in the nonet shown in Figure 15.3.2.

Figure 15.3.1: *Meson nonet with $J^P = 0^-$, called pseudoscalar mesons.*

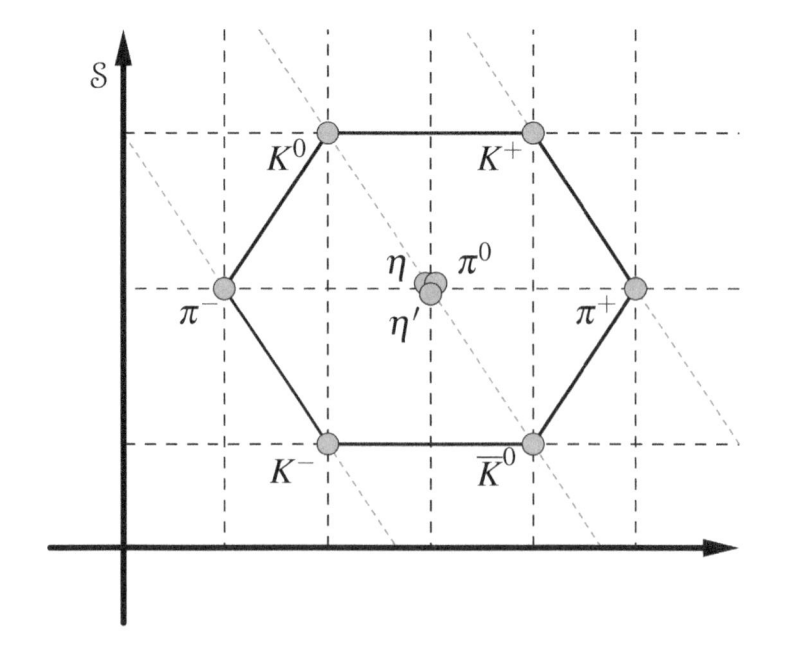

15.3 Hadrons

Figure 15.3.2: *Meson nonet with $J^P = 1^-$, called vector mesons.*

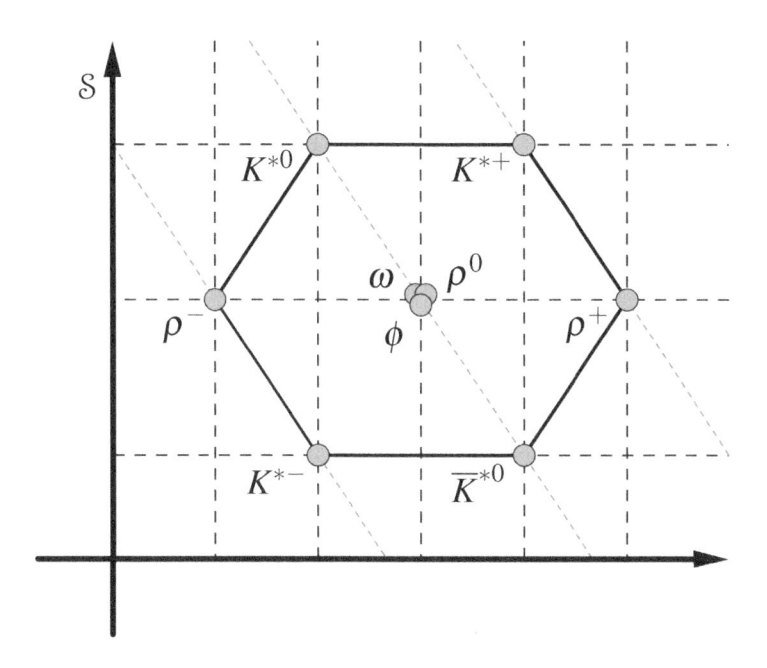

15.3.2 The Yukawa meson

Yukawa predicted that the strong nuclear force that held protons and neutrons together in the nucleus could be described by a potential, called Yukawa potential, of the type

$$\Phi(r) = \frac{g_0}{4\pi r} e^{-\frac{mc}{\hbar r}}$$

with g_0 constant and where m is the mass of the boson mediator. For the interaction radius r_0 and the mass of the boson we can write
$$r_0 \approx c\Delta t,$$
and
$$\Delta E \approx mc^2.$$
From the uncertainty principle
$$\Delta E \Delta t \approx \hbar$$
we have
$$mc\, r_0 \approx \hbar,$$
hence the mass
$$m \approx \frac{\hbar}{cr_0}.$$
If we consider a short range interaction of
$$r_0 \approx 1-2 \text{ fm},$$
we obtain
$$mc^2 \approx 100-200 \text{ MeV}.$$

15.3.3 Baryons

Baryons are non-elementary particles and are composed by three valence quarks (or three antiquarks). They have semi-integer spin and are therefore fermions. In the quark model they are represented by qqq or $\overline{qqq}$. They belong to the multiplets that can be obtained by the product of the representations

$$3 \otimes 3 \otimes 3 = 1_A \oplus 8_{MA} \oplus 8_{MS} \oplus 10_S,$$

i.e. they belong to an octet, a singlet and a decuplet. The baryons with $J^P = 1/2^+$ can be represented in the octet shown in Figure 15.3.3, while those with $J^P = 3/2^+$ can be represented in the diagram shown in Figure 15.3.4.

15.4 Nucleons

Nucleons are the constituents of the atomic nucleus, they are the proton p and the neutron n. Both are baryons and their quark composition is

$$p = uud, \qquad n = udd.$$

The electrical charge of the proton is globally $+1$ (in units of the elementary charge e) and that of the neutron is 0, being

Figure 15.3.3: *Baryon octet with $J^P = 1/2^+$.*

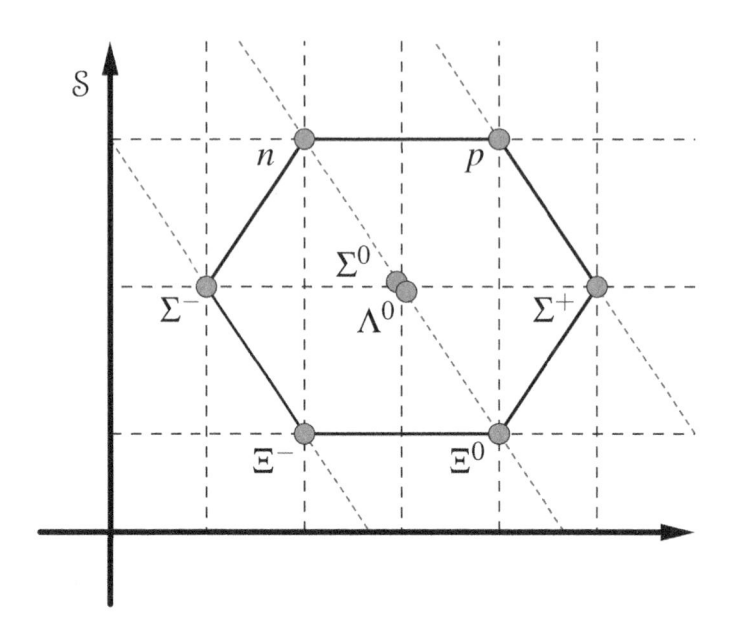

a neutral particle. Their main properties are in reported in Table 15.4.1. They are held together in the nucleus by the strong nuclear force. The mean life of the free proton can be

Table 15.4.1: *Nucleons: charge, mass and mean life.*

name and symbol	Q (e)	m (MeV)	τ (s)
proton p	+1	938.27	∞
neutron n	0	939.56	887

considered infinite, in fact it cannot decay due to the conser-

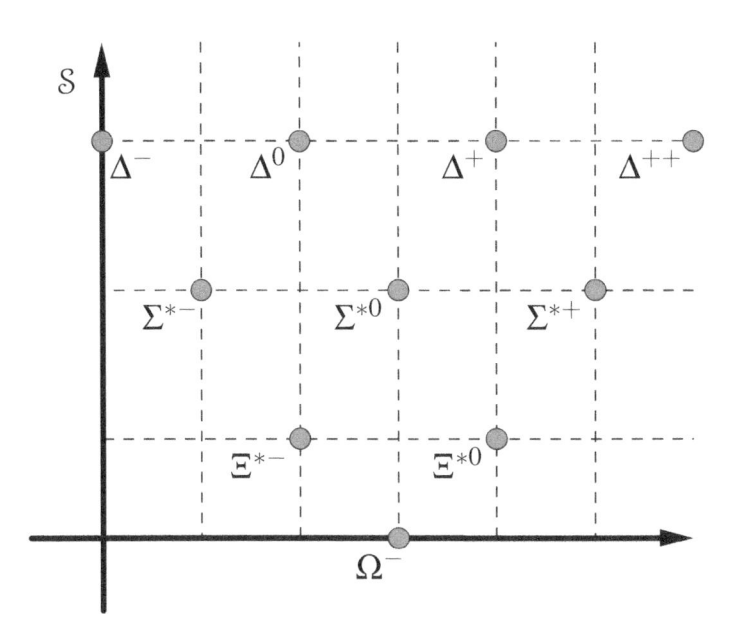

Figure 15.3.4: *Baryon decuplet with $J^P = 3/2^+$.*

vation of baryon number, being the lightest baryon.

15.5 Cosmic rays

The particles that come from space and collide with our atmosphere are called cosmic rays. Cosmic rays are mainly composed of protons (85%), α particles (12%) or electrons (2%). The α particle is a bound state of two protons and two neutrons (a helium nucleus ^4_2He). The particles in the cosmic rays can have energies variable in a large range and their flux decreases as their energy increases. The collisions of cosmic rays with the Earth's atmosphere produces other particles which decay or interact with other particles.

15.6 The pion

The pion is a particle composed by a quark and an antiquark and it is therefore a meson. Its properties are summarized in Table 15.6.1. Generally the charged pion decays weakly

Table 15.6.1: *The pion: charge, mass and mean life.*

name and symbol	Q (e)	m (MeV)	τ (μs)
pion π^+	$+1$	139.57	$2.6 \cdot 10^{-2}$
antipion π^-	-1	139.57	$2.6 \cdot 10^{-2}$
pion $\pi^0 = \overline{\pi}^0$	0	134.98	$8.4 \cdot 10^{-11}$

into muons and muon neutrinos or into electrons and electron neutrinos, while the neutral pion decays electromagnetically into photons and lepton-antilepton pairs. These decays are

$$\pi^+ \to \mu^+ + \nu_\mu,$$
$$\pi^- \to \mu^- + \overline{\nu}_\mu,$$
$$\pi^+ \to \mu^+ + \nu_\mu,$$
$$\pi^- \to \mu^- + \overline{\nu}_\mu,$$
$$\pi^0 \to \gamma + \gamma,$$
$$\pi^0 \to \gamma + e^+ + e^-.$$

15.7 The muon

The muon is an elementary particle and it is a lepton. Its properties are summarized in Table 15.7.1.

Table 15.7.1: *The muon: charge, mass and mean life.*

name and symbol	Q (e)	m (MeV)	τ (μs)
muon μ^-	-1	105.66	2.2
antimuon μ^+	$+1$	105.66	2.2

15.8 Particles with strangeness

Abnormal behavior particles were observed around the middle of the last century. For example, some of these were always produced in pairs, moreover some particles that should

Table 15.7.2: *Kaons: charge, mass, mean life and strangeness.*

symbol	Q (e)	m (MeV)	τ (ps)	S
K^+	+1	494	12	+1
K^-	−1	494	12	−1
K^0	0	498	n.d.	+1
$\overline{K}^0$	0	498	n.d.	−1

Table 15.7.3: *Hyperons: charge, mass, mean life and strangeness.*

symbol	Q (e)	m (MeV)	τ (ps)	S
Λ	0	1116	263	−1
Σ^+	+1	1189	80	−1
Σ^-	−1	1197	148	−1
Σ^0	0	1193	$7.4 \cdot 10^{-8}$	−1
Ξ^-	−1	1321	164	−2
Ξ^0	0	1315	290	−2

decay through the strong interaction had a mean life too long. These particles were named "strange particles". A new quantum number, the strangeness, was introduced, which was conserved in electromagnetic and strong interactions. Some particles were produced in pairs to preserve the strangeness.

15.8 Particles with strangeness

15.8.1 Kaons

Some of these particles are called kaons, symbol K. They are mesons and their properties are reported in Table 15.7.2.

15.8.2 Hyperons

Other particles with strangeness are the so-called hyperons. Their properties are shown in Table 15.7.3.

Chapter 16

Energy loss

16.1 Ionization energy loss

A relativistic charged particle, with a mass much greater than that of the electron m_e, through the matter interacts with atomic electrons and loses energy. It is possible to detect the ion-electron pairs produced along the path of the particle. The mean energy loss by ionization per distance travelled of a particle with electric charge z can be approximated by the Bethe-Bloch formula (Bethe 1930)

$$\frac{dE}{dx} = K\frac{\rho Z}{A}\frac{z^2}{\beta^2}\ln\left(\frac{2c^2\gamma^2 m_e - I}{I}\beta^2\right), \qquad (16.1.1)$$

where ρ is the density of the material, Z and A are its atomic and mass numbers, respectively, K is the constant

$$K = \frac{4\pi(\alpha\hbar c)^2 N_A (10^3 \text{ kg})}{m_e c^2} = 30.7 \text{ keV m}^2 \text{ kg}^{-1}$$

and I is a mean ionization potential. Its value can be calculated using the formula

$$I \approx 12 \cdot Z,$$

valid if $Z > 20$.

16.2 Electron energy loss

In the case of an electron (or positron), the Eq. (16.1.1) can no longer be used. In fact, since the electron has a small mass, m_e, it loses energy not only by ionization but also by bremsstrahlung in the Coulomb field of the nuclei of the matter. By denoting with N a nucleus we can write the processes

$$e^+ + N \rightarrow e^+ + N + \gamma,$$

or

$$e^- + N \rightarrow e^- + N + \gamma.$$

The probability, for a charged particle, of emitting a photon is proportional to its squared acceleration and therefore this phenomenon is greater near a nucleus and in materials where Z is large. Given a material we define its radiation length L the distance by which

$$\frac{dE}{E} = -\frac{dx}{L}.$$

16.3 Photon energy loss

At energies of few tens of eV a photon loses energy mainly by photoelectric effect on atomic electrons. When the energy reaches few keV the Compton effect predominates. With energy greater than 1 MeV the dominant channel becomes the e^+e^- pairs creation that can be written as

$$\gamma + N \to e^- + e^+ + N,$$

where N is a nucleus.

16.4 Hadron energy loss

High-energy hadrons through matter do not lose energy only by ionization, in fact they can interact with the nuclei of the atoms of the material due to the strong interaction. The collision length λ_0 of a material can be defined as the distance beyond which a neutron beam is attenuated by a factor $1/e$ in the material, being e the Neper number.

Chapter 17

Quantum numbers and symmetries

17.1 The strangeness

The strangeness is a quantum number. The so-called strange particles are those for which $S \neq 0$ and contain at least a strange quark s (of strangeness $+1$) or its antiquark $\bar{s}$ (of strangeness -1). The first strange particles revealed were the K mesons, called kaons, and the hyperons. The interactions that conserve the strangeness are the strong nuclear force and the electromagnetic one.

17.2 The parity

The parity operator $\hat{P}$ make the inversion of the three spatial Cartesian axes. The parity operation reverses the coordinates and leaves the time unchanged, i.e.

$$\vec{r} \to -\vec{r}, \qquad t \to t,$$

therefore for the momentum, the angular momentum and the spin we have

$$\vec{p} \to -\vec{p},$$
$$\vec{r} \times \vec{p} \to \vec{r} \times \vec{p},$$
$$\vec{s} \to \vec{s}.$$

A state has a defined parity P only if it is an eigenstate of the parity operator $\hat{P}$. A single particle can be in an eigenstate of $\hat{P}$ only if it is at rest, in this case, P is called intrinsic parity and it can be a real number or an imaginary number. For bosons and antibosons the intrinsic parity is the same, while for fermions and antifermions we have

$$P_f P_{\bar{f}} = -1.$$

17.3 Parity of the photon

The photon has parity $P = -1$ and $J = 1$, so that in notation J^P we write

$$J^P = 1^-.$$

17.4 Parity of a two-particle system

The parity of a system, with orbital angular momentum l, of two particles with intrinsic parity P_1 and P_2 is given by

$$P_{(2\,particles)} = P_1 P_2 (-1)^l.$$

For a boson-antiboson pair, $b\bar{b}$, since $P_b = P_{\bar{b}}$, we have

$$P_{(b\bar{b})} = (-1)^l,$$

while, for a fermion-antifermion pair, $f\bar{f}$, being $P_f P_{\bar{f}} = -1$, it follows that

$$P_{(f\bar{f})} = (-1)^{l+1}.$$

17.5 Charge conjugation

The operator of particle-antiparticle conjugation, or charge conjugation, $\hat{C}$, acting on a one-particle state, changes the particle with its antiparticle. Applying twice the operator $\hat{C}$ we recover the initial state and therefore the possible eigenvalues are $C = \pm 1$. The application of $\hat{C}$ changes the charge of a particle, therefore to be an eigenstate of $\hat{C}$ a particle must be neutral. The eigenstates of $\hat{C}$ are particles that coincide with their antiparticle, such as the photon and the neutral pion π^0.

17.6 Charge conjugation of the photon

For the photon we have

$$\hat{C}|\gamma\rangle = -|\gamma\rangle.$$

This can be deduced from the correspondence of the potential classical vector A with the photon. For a state of n photons we have

$$\hat{C}|n\gamma\rangle = (-1)^n |n\gamma\rangle.$$

17.7 Charge conjugation of the pion

The electromagnetic interaction conserves the charge conjugation and therefore from the decay

$$\pi^0 \to \gamma + \gamma$$

we can write

$$\hat{C}|\pi^0\rangle = +|\pi^0\rangle.$$

The charged pions are not eigenstates of $\hat{C}$, in fact

$$\hat{C}|\pi^-\rangle = |\pi^+\rangle$$

and

$$\hat{C}|\pi^+\rangle = |\pi^-\rangle.$$

17.8 Time reversal

The time reversal operator $\hat{T}$ reverses the time.

17.9 CPT theorem

The *CPT* theorem states that if an interacting field theory is invariant under the proper Lorentz group then it will also be invariant under the combination of the three operators $\hat{C}$, $\hat{P}$

and $\hat{T}$ in any order.

17.10 Baryon number

The baryon number $\mathcal{B}$ of a state is defined as the number of baryons minus the number of antibaryons

$$\mathcal{B} = \#(\text{baryons}) - \#(\text{antibaryons}).$$

All known interactions conserve the baryon number.

17.11 Lepton number

The total lepton number, or simply leptonic number, $\mathcal{L}$ of a state is defined as the number of leptons minus the number of antileptons

$$\mathcal{L} = \#(\text{leptons}) - \#(\text{antileptons}).$$

We can also define the three partial lepton numbers: the electron number $\mathcal{L}_e$, the muon number $\mathcal{L}_\mu$ and the tau number $\mathcal{L}_\tau$. Their definitions are

$$\mathcal{L}_e = \left(\#(e^-) + \#(\nu_e)\right)$$
$$- \left(\#(e^+) + \#(\overline{\nu}_e)\right),$$

$$\mathcal{L}_\mu = \Big(\#(\mu^-) + \#(\nu_\mu)\Big)$$
$$- \Big(\#(\mu^+) + \#(\overline{\nu}_\mu)\Big),$$
$$\mathcal{L}_\tau = \Big(\#(\tau^-) + \#(\nu_\tau)\Big)$$
$$- \Big(\#(\tau^+) + \#(\overline{\nu}_\tau)\Big),$$

and we have
$$\mathcal{L} = \mathcal{L}_e + \mathcal{L}_\mu + \mathcal{L}_\tau.$$

All known interactions conserve the total lepton number.

17.12 Isospin

Nuclear forces are independent of electric charge. The proton and the neutron can be considered as two states of the same particle, the nucleon. We can introduce the quantum number for the strong nuclear force, the isotopic spin, called isospin, which follows the formal rules of an angular momentum. The nucleon isospin is therefore $I = 1/2$ and the two third components $I_3 = +1/2$ and $I_3 = -1/2$ correspond, respectively, to the proton and to the neutron which are components of an isospin doublet. All particles of a multiplet must have the same mass, the same spin and the same parity.

The mass difference between the proton and the neutron is given by electromagnetic effects, otherwise they would have the same mass. The pion has isospin 1 and therefore there are three states corresponding to the three third components $+1, -1, 0$. These are, respectively, the charged pions π^+ and π^- and the neutral pion π^0.

17.13 Hypercharge

We can introduce the hypercharge $\mathcal{Y}$ defined initially as

$$\mathcal{Y} = \mathcal{B} + \mathcal{S},$$

or as the sum of the baryonic number and strangeness. In particular for mesons, being $\mathcal{B}_{mesons} = 0$ we have

$$\mathcal{Y}_{mesons} = \mathcal{S}_{mesons}.$$

17.14 The Gell-Mann-Nishijima formula

The Gell-Mann-Nishijima formula involves the third component of the isospin I_3, the electric charge Q and the hypercharge $\mathcal{Y}$ and states that

$$Q = I_3 + \frac{\mathcal{Y}}{2}.$$

17.15 G-parity

For example, for the neutron we have

$$\underbrace{0}_{Q} = \underbrace{-1/2}_{I_3} + \frac{1}{2} \cdot \underbrace{1}_{y},$$

while for the proton

$$\underbrace{+1}_{Q} = \underbrace{+1/2}_{I_3} + \frac{1}{2} \cdot \underbrace{1}_{y}.$$

17.15 G-parity

The G-parity is defined as

$$\hat{G} = e^{-i\pi I_2} \hat{C},$$

where $\hat{C}$ is the charge conjugation operator and I_2 is the y component of the isospin. For any pion (π^+, π^-, π^0) we have

$$\hat{G}|\pi\rangle = -|\pi\rangle,$$

and therefore $G = -1$, while for a system of n_π pions

$$\hat{G}|n_\pi \pi\rangle = (-1)^{n_\pi}|n_\pi \pi\rangle$$

and
$$G = (-1)^{n_\pi}.$$

17.16 Helicity

For a fermion helicity states are defined as the eigenstates of the helicity operator
$$\frac{\vec{p} \cdot \vec{\sigma}}{2p},$$
where $\vec{p}$ is the momentum of the particle, p is its module and $\vec{\sigma}$ is the spin operator of the particle, given by the Pauli matrices. Only for massless fermions the helicity is a Lorentz invariant.

17.17 Chirality

The chirality states for a fermion are the eigenstates of the Dirac γ_5 matrix. The possible eigenvalues are ± 1 and are said to be R (right) in the case of $+1$ and L (left) in the case of -1. The two projectors of the R and L states of chirality can be defined as
$$\hat{P}_R = \frac{1}{2}(1 - \gamma_5)$$
and
$$\hat{P}_L = \frac{1}{2}(1 + \gamma_5).$$

17.17 Chirality

So the two states are

$$\psi_R = \frac{1}{2}(1-\gamma_5)\psi$$

and

$$\psi_L = \frac{1}{2}(1+\gamma_5)\psi,$$

where ψ is the solution of the Dirac equation. Only L states contribute to weak charge current interactions. Chirality is a good quantum number for a free fermion only if it is massless. Sometimes, at high energies, the mass of a fermion can be neglected and chirality can be used as a good quantum number.

Chapter 18

Scattering and decays

18.1 Reference frames

The calculation of many quantities may depend on the chosen reference frame. The transformations that move from one system to another are those of Lorentz. Some quantities are Lorentz-invariant and therefore the calculation in any reference system produces the same result. In this case, we usually choose the reference frame where the calculations are simpler. The two most commonly used reference frames are the laboratory frame and the center of mass frame. In the center of mass system the sum of all the momentum vectors, $\vec{p}_i$, of the particles is zero, by definition. The quantities calculated in the center of mass are generally indicated with an asterisk. In Figure 18.1.1 we show the two systems for two particles.

Figure 18.1.1: *Laboratory frame and center of mass frame (CM) for two particles a and b.*

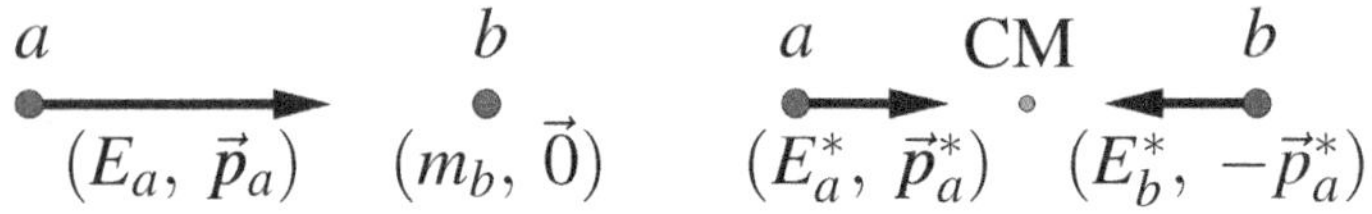

$$a \quad\quad b \quad\quad\quad a \quad CM \quad b$$
$$(E_a, \vec{p}_a) \quad (m_b, \vec{0}) \quad (E_a^*, \vec{p}_a^*) \quad (E_b^*, -\vec{p}_a^*)$$

18.2 The invariant quantity s

Given a system of N non-interacting particles, each of energy E_i and momentum $\vec{p}_i$ we can define the Lorentz invariant quantity s as the square of the invariant mass

$$s = m^2 = \left(\sum_{i=1}^{N} E_i\right)^2 - \left(\sum_{i=1}^{N} \vec{p}_i\right)^2.$$

In the center of mass frame, indicating with an asterisk the quantities referred to it, we have

$$s = \left(\sum_{i=1}^{N} E_i^*\right)^2. \quad\quad (18.2.1)$$

18.2 The invariant quantity s

From here we also see that the invariant mass of a system of N non-interacting particles represents its energy in the center of mass frame, E^*, i.e.

$$m = \sum_{i=1}^{N} E_i^*$$

Consider now a system of two free particles ($N = 2$). In the laboratory frame, by denoting with $\vec{p}_1, \vec{p}_2$ the momenta and with E_1, E_2 the energies we can write

$$s = (E_1 + E_2)^2 - (\vec{p}_1 + \vec{p}_2)^2,$$

or

$$s = E_1^2 + E_2^2 + 2E_1 E_2 - p_1^2 - p_2^2 - 2\vec{p}_1 \vec{p}_2,$$

with $p_i = |\vec{p}_i|$. Using the mass-shell relation

$$E_i^2 = m_i^2 + p_i^2,$$

we have

$$\begin{aligned}s &= m_1^2 + m_2^2 + 2E_1E_2\left(1 - \frac{\vec{p}_1\vec{p}_2}{E_1E_2}\right) \\ &= m_1^2 + m_2^2 + 2E_1E_2\left(1 - \vec{\beta}_1\vec{\beta}_2\right),\end{aligned}$$

being

$$\vec{\beta}_i = \vec{p}_i/E_i$$

In summary, the quantity s can therefore be written in the two equivalent ways

$$s = m_1^2 + m_2^2 + 2(E_1E_2 - \vec{p}_1\vec{p}_2), \qquad (18.2.2)$$

or

$$s = m_1^2 + m_2^2 + 2E_1E_2\left(1 - \vec{\beta}_1\vec{\beta}_2\right). \qquad (18.2.3)$$

18.3 Mandelstam variables

For a scattering process of the type

$$a + b \to c + d,$$

we can define three variables, called Mandelstam variables, which are Lorentz invariants and are called s, t, u. Let p_a, p_b,

p_c and p_d be the four-momenta of the four particles. The firs is the s variable, introduced previously,

$$s = m^2 = (p_a + p_b)^2 = (p_c + p_d)^2,$$

while the other are

$$t = (p_a - p_c)^2 = (p_b - p_d)^2,$$

and

$$u = (p_a - p_d)^2 = (p_b - p_c)^2.$$

We have also

$$s + t + u = m_a^2 + m_b^2 + m_c^2 + m_d^2,$$

where m_a, m_b, m_c and m_d are, respectively, the masses of the four particles.

18.4 Two-body elastic scattering

An elastic scattering between two particles, called a and b can be written as

$$a + b \to a + b,$$

where the initial and final states have the same particles. The time duration of the interaction is very small compared to the time duration of the measurements on initial and final states, therefore the particles can be considered free in both cases. Suppose that in the initial state, in the laboratory frame, the particle b, of mass m_b is at rest, while the particle a, of mass m_a, has momentum $\vec{p}_a$, energy E_a and is directed against b. The energy of the b particle in the initial state is

$$E_b = \sqrt{p_b^2 + m_b^2} = m_b$$

and the variable s, using the Eq. (18.2.2) with $\vec{p}_a \vec{p}_b = 0$ and $E_b = m_b$, can be written as

$$s = m_a^2 + m_b^2 + 2m_b E_a.$$

For the energy and momentum conservation, s must be the same in both the initial and final states. We observe that if the particle a has an high energy, i.e. if $E_a \gg \max(m_a, m_b)$, then

$$s \approx 2m_b E_a. \qquad (18.4.1)$$

18.4 Two-body elastic scattering

If we perform the same calculation in the center of mass frame, where

$$\vec{p}_a^* = -\vec{p}_b^*$$

and

$$p_a^* = p_b^* = p^*,$$

using the Eq. (18.2.1), we obtain

$$s = (E_a^* + E_b^*)^2 = {E_a^*}^2 + {E_b^*}^2 + 2 E_a^* E_b^*, \qquad (18.4.2)$$

in this case we have, from the mass-shell relation,

$${E_a^*}^2 = {p_a^*}^2 + m_a^2 = {p^*}^2 + m_a^2,$$

and

$${E_b^*}^2 = {p_b^*}^2 + m_b^2 = {p^*}^2 + m_b^2,$$

moreover

$$E_a^* E_b^* = \sqrt{({p^*}^2 + m_a^2)({p^*}^2 + m_b^2)},$$

from which

$$\begin{aligned}s =\ & 2p^{*2}+m_a^2+m_b^2 \\ & + 2\sqrt{(p^{*2}+m_a^2)(p^{*2}+m_b^2)}.\end{aligned}$$
(18.4.3)

If we assume that $m_i \ll E_i^*$ for $i = a, b$, from the above mass-shell relations we have

$$E_a^* \approx p^*,$$

and

$$E_b^* \approx p^*,$$

therefore

$$E^* := E_a^* \approx E_b^* \approx p^*.$$

Under these considerations the Eq. (18.4.2), or equivalently, the Eq. (18.4.3), becomes

$$s \approx (2E^*)^2.$$

18.5 Fermi's golden rule

Summarizing, if a scattering is performed on a fixed target, that is a particle a with energy E_a hits a particle b at rest, then the energy of the center of mass available is, from Eq. (18.4.1),

$$m = \sqrt{s} \approx \sqrt{2m_b}\sqrt{E_a},$$

proportional to the square root of the energy of a. On the other hand, if you perform a scattering with particles of energy E^* and opposite momentum (collision of particles), the calculation made in the center of mass frame gives the available energy

$$m = \sqrt{s} \approx 2E^*$$

which is, on the other hand, proportional to the energy of each particle.

18.5 Fermi's golden rule

Suppose we have a system in an initial eigenstate $|i\rangle$ of a Hamiltonian H_0. Suppose we introduce a perturbation Hamiltonian H_p in addition to H_0, the probability per unit time that for the transition from initial state to final one $|f\rangle$ is given, at the first perturbative order, by

$$W_{i \to f} = 2\pi \left|\langle f|H_p|i\rangle\right|^2 \rho(E),$$

where $\rho(E)$ is the density of the final state at the final energy E, also called phase space. We can define also the transition matrix

$$\mathcal{M}_{fi} := \langle f|H_p|i\rangle.$$

The phase space for a transition with n particles in the final state is given by

$$\rho_n(E) = (2\pi)^4 \int \prod_{i=1}^{n} \frac{d^3 p_i}{2E_i (2\pi)^3} \cdot \delta\left(\sum_{i=1}^{n} E_i - E\right)$$
$$\cdot \delta^3\left(\sum_{i=1}^{n} \vec{p}_i - \vec{P}\right).$$

18.6 Cross section

Suppose we have a particle beam that hits a material. Let v_p be the module of the beam velocity and n_p its density. Let S_b be the surface of the target exposed perpendicular to the beam, L_b its thickness and N_b the number of target particles (scattering centers). The probability of a particle hitting the

18.6 Cross section

target is given by

$$\mathcal{P} = \frac{\text{effective impact area}}{\text{material area}} = \frac{N_b \sigma}{S_b}$$

where σ is the effective impact area with a single target, the so-called cross section. If N_p particles arrive to the material, the number of interacting particles, called event numbers, is given by

$$N_e = N_p \mathcal{P} = \frac{N_b N_p \sigma}{S_b}.$$

The flux of arriving particles is

$$\Phi_p = \frac{dN_p}{dSdt} = \frac{v_p dN_p}{dSdx_p} = \frac{n_p v_p dV}{dSdx_p} = \frac{n_p v_p dSdx_p}{dSdx_p} = n_p v_p.$$

We can write the number of particles N_p arriving on the target material of area S_b in a time Δt as

$$N_p = \Phi_p S_b \Delta t.$$

The number of events is

$$N_e = \frac{N_b \Phi_p S_b \sigma \Delta t}{S_b} = N_b \Phi_p \sigma \Delta t.$$

18. Scattering and decays

We define the total number of events (interactions) per unit time as

$$R = \frac{N_e}{\Delta t},$$

from which the cross section

$$\sigma = \frac{R}{N_b \Phi_p}. \qquad (18.6.1)$$

We can write

$$R = N_b W,$$

where W is the rate of particles in the material. Generally W is given by the Fermi's golden rule, therefore

$$\sigma = \frac{W}{\Phi_p}. \qquad (18.6.2)$$

We can write

$$\sigma = \int \frac{d\sigma}{d\Omega} d\Omega,$$

where the quantity

$$\frac{d\sigma}{d\Omega}$$

18.6 Cross section

is the differential cross section and

$$d\Omega = \cos\theta \, d\theta \, d\phi$$

is the infinitesimal solid angle.

18.6.1 Beam intensity reduction

To calculate the beam intensity reduction we call with $I(x)$ the intensity of the beam in the x position in the passage through the material. Let I_0 be the intensity of the initial beam and dR the number of total interactions per unit time for a length dx in the material. The variation of the beam intensity can be written as

$$dI(x) = -dR. \tag{18.6.3}$$

For the target surface S_b (orthogonal to the beam) the flux will be

$$\Phi_p = \frac{I(x)}{S_b}.$$

From the Eq. (18.6.1), using the density of the target material, n_b, we have

$$dR = \sigma \Phi_p dN_b = \sigma \Phi_p S_b n_b dx = \sigma I(x) n_b dx,$$

from which, using the Eq. (18.6.3),

$$dI(x) = -\sigma I(x) n_b dx,$$

or

$$\frac{dI(x)}{I(x)} = -\sigma n_b dx.$$

Finally we have

$$I(x) = I_0 e^{-n_b \sigma x}$$

and we can define an absorption length

$$L = \frac{1}{n_b \sigma},$$

from which

$$I(x) = I_0 e^{-x/L}.$$

The absorption length L indicates the distance at which the beam intensity is reduced by a factor $1/e$, where e is the

18.6 Cross section

Neper number.

18.6.2 Luminosity

We define the luminosity $\mathcal{L}$ as the ratio between the number of interactions per unit time R and the cross section. In formulas

$$\mathcal{L} = \frac{R}{\sigma}.$$

From the Eq. (18.6.1) we also have

$$\mathcal{L} = N_b \Phi_p.$$

18.6.3 Two-body cross section

In the case of the scattering where we have two particles a and b in the initial state and n particles in the final state, the calculation of the cross section given in Eq. (18.6.2), using the Fermi's golden rule for W, is

$$\sigma = \frac{(2\pi)^4}{4E_a E_b |\vec{\beta}_a - \vec{\beta}_b|} \int |\mathcal{M}_{fi}|^2 \cdot \prod_{i=1}^{n} \frac{d^3 p_i}{2E_i (2\pi)^3}$$
$$\cdot \delta\left(\sum_{i=1}^{n} E_i - E\right) \delta^3\left(\sum_{i=1}^{n} \vec{p}_i - \vec{P}\right),$$

where E_a, E_b are the initial energies of the two particles a and b and $\vec{\beta}_a, \vec{\beta}_b$ are their velocities (in natural units).

18.7 Decays

In a decay we have a particle in the initial state which decays into two or more particles in the final states

$$a \to b+c+\cdots.$$

If a particle can decay in several ways, each of them is called a decay channel. The partial decay rate, or partial width, of a in one channel, for example the final state b, c, is denoted with Γ_{bc}. The sum of all partial widths gives the total width Γ for the particle a which is also the reciprocal of its mean life

$$\Gamma = \frac{1}{\tau},$$

in fact from the uncertainty principle

$$\Gamma \tau \approx \hbar = 1.$$

18.7 Decays

We call branching ratio (BR) of a in b, c the ratio

$$R_{bc} = \frac{\Gamma_{bc}}{\Gamma}.$$

For a decay, using the Fermi's golden rule, the probability of transition per unit time to a final state of n particles is given by

$$\Gamma_{i,f} = \frac{(2\pi)^4}{2E} \int |\mathcal{M}_{fi}|^2 \prod_{i=1}^{n} \frac{d^3 p_i}{2E_i (2\pi)^3}$$
$$\cdot \delta\left(\sum_{i=1}^{n} E_i - E\right) \delta^3\left(\sum_{i=1}^{n} \vec{p}_i - \vec{P}\right).$$

where E is the initial energy of the a particle that decays. For a two-body decay, i.e.

$$a \to b + c,$$

the explicit calculation in the center of mass frame gives

$$\Gamma_{a,bc} = \frac{p_f}{32\pi^2 m^2} \int |\mathcal{M}_{a,bc}|^2 d\Omega_f,$$

with $p_f = p_b = p_c$ the modulus of the momentum and m the mass of the particle a. This can also be written as

$$\Gamma_{a,bc} = \frac{p_f}{8\pi m^2} \overline{|\mathcal{M}_{a,bc}|^2}.$$

Part III

Theoretical Physics

Chapter 19

Introduction

In this third part we introduce the basic concepts of the theoretical physics. The topics covered are the following:
- Lagrangian field theory;
- Hamiltonian field theory;
- symmetries;
- the gauge invariance;
- the Klein-Gordon field;
- the Dirac field;
- an introduction to the QED.

Chapter 20

Lagrangian and Hamiltonian

20.1 Lagrangian field theory

Generally the Lagrangian can be written as a spatial integral of the Lagrangian density in the following way

$$L = \int d^3x \, \mathcal{L},$$

so that the action can take the form

$$S = \int dt \, L = \int d^4x \, \mathcal{L}.$$

20. Lagrangian and Hamiltonian

In this book we will always refer to the Lagrangian density, being a Lorentz invariant quantity. In field theory, the Lagrangian density for a field ϕ is a function of the field itself and its derivatives and we can write

$$\mathcal{L} = \mathcal{L}(\phi, \partial_\mu \phi).$$

The principle of least action is written as

$$\begin{aligned}\delta S &= \int d^4x \left[\frac{\partial \mathcal{L}}{\partial \phi} \delta\phi + \frac{\partial \mathcal{L}}{\partial(\partial_\mu \phi)} \delta(\partial_\mu \phi) \right] \\ &= \int d^4x \left[\frac{\partial \mathcal{L}}{\partial \phi} \delta\phi - \left(\partial_\mu \frac{\partial \mathcal{L}}{\partial(\partial_\mu \phi)} \right) \delta\phi \right. \\ &\quad + \left. \partial_\mu \left(\frac{\partial \mathcal{L}}{\partial(\partial_\mu \phi)} \delta\phi \right) \right] = 0, \end{aligned}$$

therefore

$$\partial_\mu \left(\frac{\partial \mathcal{L}}{\partial(\partial_\mu \phi)} \delta\phi \right) = \left(\partial_\mu \frac{\partial \mathcal{L}}{\partial(\partial_\mu \phi)} \right) \delta\phi + \frac{\partial \mathcal{L}}{\partial(\partial_\mu \phi)} \delta(\partial_\mu \phi). \quad (20.1.1)$$

Neglecting the last term in the integral which would produce a null surface term we get

$$\delta S = \int d^4x \left[\frac{\partial \mathcal{L}}{\partial \phi} - \left(\partial_\mu \frac{\partial \mathcal{L}}{\partial (\partial_\mu \phi)} \right) \right] \delta \phi = 0.$$

The variation of the action must be zero for arbitrary values of the variation of the field, i.e. $\delta \phi$, and the equations of motion are

$$\partial_\mu \frac{\partial \mathcal{L}}{\partial (\partial_\mu \phi)} = \frac{\partial \mathcal{L}}{\partial \phi}. \qquad (20.1.2)$$

20.2 Hamiltonian field theory

Given a Lagrangian density

$$\mathcal{L} = \mathcal{L}(\phi_i, \partial_\mu \phi_i),$$

we define the conjugated momentum densities to the fields ϕ_i the following

$$\pi_i(x) = \frac{\partial \mathcal{L}}{\partial \dot{\phi}_i(x)}. \qquad (20.2.1)$$

The Hamiltonian can be written in terms of a Hamiltonian density in the following way

$$H = \int d^3x \, \mathcal{H}, \qquad (20.2.2)$$

where, in terms of the conjugated momenta $\pi_i(x)$,

$$\mathcal{H} = \sum_i \pi_i(x) \dot{\phi}_i(x) - \mathcal{L}. \qquad (20.2.3)$$

Chapter 21

Symmetries and gauge invariance

21.1 Symmetries and conservation laws

We can define symmetry a transformation of the fields that leaves the equation of motion unchanged. This happens if the transformation modifies the Lagrangian density by adding a four-divergence of an arbitrary function (surface term that does not modify the equations of motion). For example the transformation

$$\phi \to \phi' = \phi + \delta\phi$$

21. Symmetries and gauge invariance

is a symmetry if there is a function $\tilde{J}^\mu(x)$ such that the variation of the Lagrangian density can be written as

$$\mathcal{L} \to \mathcal{L}' = \mathcal{L} + \delta\mathcal{L}$$

with

$$\delta\mathcal{L} = \partial_\mu \tilde{J}^\mu. \qquad (21.1.1)$$

We calculate, using the Eq. (20.1.1),

$$\begin{aligned}
\delta\mathcal{L} &= \frac{\partial \mathcal{L}}{\partial \phi}\delta\phi + \frac{\partial \mathcal{L}}{\partial(\partial_\mu \phi)}\delta(\partial_\mu \phi) \\
&= \frac{\partial \mathcal{L}}{\partial \phi}\delta\phi + \partial_\mu\left(\frac{\partial \mathcal{L}}{\partial(\partial_\mu \phi)}\delta\phi\right) - \left(\partial_\mu \frac{\partial \mathcal{L}}{\partial(\partial_\mu \phi)}\right)\delta\phi \\
&= \partial_\mu\left(\frac{\partial \mathcal{L}}{\partial(\partial_\mu \phi)}\delta\phi\right) + \left(\frac{\partial \mathcal{L}}{\partial \phi} - \partial_\mu \frac{\partial \mathcal{L}}{\partial(\partial_\mu \phi)}\right)\delta\phi \\
&= \partial_\mu\left(\frac{\partial \mathcal{L}}{\partial(\partial_\mu \phi)}\delta\phi\right),
\end{aligned}$$

in fact the last term is zero thanks to the Eq. (20.1.2). Moreover from the Eq. (21.1.1) we can write

$$\partial_\mu \tilde{J}^\mu = \partial_\mu \frac{\partial \mathcal{L}}{\partial(\partial_\mu \phi)}\delta\phi,$$

from which
$$\tilde{J}^\mu = \frac{\partial \mathcal{L}}{\partial(\partial_\mu \phi)}\delta\phi - J^\mu,$$

where the current J^μ is conserved

$$\partial_\mu J^\mu = 0$$

and

$$J^\mu = \frac{\partial \mathcal{L}}{\partial(\partial_\mu \phi)}\delta\phi - \tilde{J}^\mu.$$

21.2 Gauge invariance

The interaction Lagrangian in field theories can emerge naturally considering the gauge invariance. Given a particle field $\psi(x)$ suppose we want to perform a gauge transformation given by the unitary and local operator $U(x)$ of a certain group which acts as

$$\psi(x) \to \psi'(x) = U(x)\psi(x).$$

We observe immediately that

$$\begin{aligned}\left(\partial_\mu \psi(x)\right)' &= \partial_\mu \psi'(x) = \partial_\mu\left(U(x)\psi(x)\right) \\ &= \left(\partial_\mu U(x)\right)\psi(x) + U(x)\partial_\mu \psi(x),\end{aligned}$$

which is different from

$$U(x)\left(\partial_\mu \psi(x)\right).$$

Therefore the term $\partial_\mu \psi(x)$ is not invariant under the gauge transformation given by $U(x)$, and the same would be a Lagrangian which contains the derivative of the fields. To ensure the gauge invariance, we have to introduce the so-called covariant derivative

$$D_\mu \equiv \partial_\mu + \Gamma_\mu(x), \qquad (21.2.1)$$

which is written in terms of the so-called connection $\Gamma_\mu(x)$, such that

$$\left(D_\mu \psi(x)\right)' = U(x)\left(D_\mu \psi(x)\right).$$

21.2 Gauge invariance

The first member of the equation becomes

$$\begin{aligned}\left(D_\mu \psi(x)\right)' &= \left(\partial_\mu + \Gamma_\mu(x)\right)' \psi'(x) = \left(\partial_\mu + \Gamma'_\mu(x)\right) \\ &\quad \cdot \left(U(x)\psi(x)\right) = \left(\partial_\mu U(x)\right)\psi(x) \\ &\quad + U(x)\partial_\mu \psi(x) + \Gamma'_\mu(x)U(x)\psi(x),\end{aligned}$$

while the second member is

$$\begin{aligned}U(x)\left(D_\mu \psi(x)\right) &= U(x)\left(\partial_\mu \psi(x) + \Gamma_\mu(x)\psi(x)\right) \\ &= U(x)\partial_\mu \psi(x) + U(x)\Gamma_\mu(x)\psi(x)\end{aligned}$$

and therefore from the equation we obtain

$$U(x)\Gamma_\mu(x)\psi(x) = \left(\partial_\mu U(x)\right)\psi(x) + \Gamma'_\mu(x)U(x)\psi(x).$$

This relation must be an identity for each field $\psi(x)$, so we obtain the operator identity

$$U(x)\Gamma_\mu(x) = \partial_\mu U(x) + \Gamma'_\mu(x)U(x),$$

moreover we can multiply from right by $U^{-1}(x)$ and write

$$\Gamma'_\mu(x) = U(x)\Gamma_\mu(x)U^{-1}(x) - \left(\partial_\mu U(x)\right)U^{-1}(x).$$

21. Symmetries and gauge invariance

Since $U(x)$ is unitary, from

$$U(x)U^\dagger(x) = 1,$$

we obtain

$$\Gamma'_\mu(x) = U(x)\Gamma_\mu(x)U^\dagger(x) - \left(\partial_\mu U(x)\right)U^\dagger(x). \quad (21.2.2)$$

This expression can also be written as

$$\Gamma'_\mu(x) = U(x)\Gamma_\mu(x)U^\dagger(x) + U(x)\partial_\mu U^\dagger(x),$$

in fact

$$\partial_\mu\left(U(x)\right)U^\dagger(x) + U(x)\partial_\mu U^\dagger(x) = \partial_\mu\left(U(x)U^\dagger(x)\right)$$
$$= \partial_\mu 1 = 0.$$

In the case of electromagnetism the group of transformations is $U(1)$ (abelian) therefore $U(x) \in U(1)$ unitary with[1]

$$U(x) = e^{-ie\Phi(x)}, \quad \Phi(x) \in \mathbb{R}$$

---
[1] it would be $U(x) = e^{iq\Phi(x)}$ but $q = -e$ for the electron, $e > 0$, representative of the fermion field e^+e^-.

21.2 Gauge invariance

and the connection $\Gamma_\mu(x)$ in this case is

$$\Gamma_\mu(x) = -ieA_\mu(x). \qquad (21.2.3)$$

Using the Eq. (21.2.2) we have

$$\begin{aligned}\Gamma'_\mu(x) &= e^{-ie\Phi(x)}\Gamma_\mu(x)e^{ie\Phi(x)} - \left(\partial_\mu e^{-ie\Phi(x)}\right)e^{ie\Phi(x)} \\ &= \Gamma_\mu(x) + ie\left(\partial_\mu\Phi(x)\right)e^{-ie\Phi(x)}e^{ie\Phi(x)} \\ &= -ieA_\mu(x) + ie\left(\partial_\mu\Phi(x)\right).\end{aligned}$$

Finally, from $\Gamma'_\mu(x) = -ieA'_\mu(x)$,

$$A'_\mu(x) = A_\mu(x) - \partial_\mu\Phi(x).$$

This is the gauge transformation of the photon field which is defined less than an additive four-divergence which does not modify the equations of motion. In fact, to derive the equations of motion in field theory, starting from the functional variation of the action

$$\delta S = \delta \int dt\, L(x) = \delta \int d^4x\, \mathcal{L}(x),$$

the additional term would be a surface term which provides a zero contribution. In conclusion, the request that the Lagrangian of a field of charged particles, as in QED, is (local) gauge invariant, automatically implies the introduction of an interaction term (in this case the interaction is with the photon field, which is the mediator of the electromagnetic interaction).

Chapter 22

The Klein-Gordon field

22.1 Klein-Gordon equation

The Klein-Gordon equation for a free particle in relativistic quantum mechanics can be obtained starting from the mass-shell relation

$$E^2 = \vec{p}^{\,2} + m^2.$$

In fact starting from

$$(\hat{\vec{p}}^{\,2} + m^2)\phi = \hat{E}^2 \phi$$

and replacing the operators

$$\hat{E} = i\partial^0, \quad \hat{p}^i = -i\partial^i,$$

we obtain

$$-\partial^0\partial^0\phi = -\partial^i\partial^i\phi + m^2\phi,$$

from which

$$(\partial^\mu\partial_\mu + m^2)\phi = 0,$$

or

$$(\Box + m^2)\phi = 0,$$

where

$$\Box \equiv \partial_\mu\partial^\mu.$$

22.2 Klein-Gordon Lagrangian

A Lagrangian density describing a real ϕ scalar field, using Lorentz invariance, can only have the form

$$\mathcal{L}(\phi, \partial_\mu\phi) = a\,\partial_\mu\phi\partial^\mu\phi + b\,\phi^2,$$

22.2 Klein-Gordon Lagrangian

where a, b are constant. The Klein-Gordon Lagrangian density for a non-interacting field is

$$\mathcal{L} = \frac{1}{2}\partial_\mu \phi \, \partial^\mu \phi - \frac{1}{2}m^2 \phi^2. \quad (22.2.1)$$

Using the the equations of motion shown in Eq. (20.1.2) we obtain

$$\frac{\partial \mathcal{L}}{\partial \phi} = -\frac{1}{2}m^2 \frac{\partial}{\partial \phi}\phi^2 = -m^2 \phi,$$

and

$$\begin{aligned}
\frac{\partial \mathcal{L}}{\partial(\partial_\mu \phi)} &= \frac{1}{2}\frac{\partial}{\partial(\partial_\mu \phi)}(\partial_\mu \phi \, \partial^\mu \phi) \\
&= \frac{1}{2}\frac{\partial}{\partial(\partial_\mu \phi)}(\partial_\mu \phi \, \eta^{\mu\nu} \partial_\nu \phi) \\
&= \frac{1}{2}\left(\partial^\mu \phi + \eta^{\mu\nu}\partial_\mu \phi \frac{\partial}{\partial(\partial_\mu \phi)}(\partial_\nu \phi)\right) \\
&= \frac{1}{2}(\partial^\mu \phi + \eta^{\mu\nu}\partial_\mu \phi \, \delta^\mu_\nu) \\
&= \frac{1}{2}(\partial^\mu \phi + \eta^{\mu\nu}\partial_\nu \phi) \\
&= \frac{1}{2}(\partial^\mu \phi + \partial^\mu \phi) = \partial^\mu \phi.
\end{aligned}$$

The equations of motion for the Klein-Gordon field are, using the Eq. (20.1.2),

$$\partial_\mu \partial^\mu \phi = -m^2 \phi,$$

or also

$$(\partial^\mu \partial_\mu + m^2) \phi = 0.$$

22.3 Klein-Gordon Hamiltonian

The conjugated momentum to ϕ, according to the definition given in Eq. (20.2.1), is

$$\begin{aligned}\pi(x) &= \frac{\partial \mathcal{L}}{\partial \dot{\phi}(x)} = \frac{1}{2}\frac{\partial}{\partial \dot{\phi}}(\partial_\mu \phi \, \partial^\mu \phi) = \frac{1}{2}\frac{\partial}{\partial \dot{\phi}}(\partial_0 \phi \, \partial^0 \phi) \\ &= \frac{1}{2}\frac{\partial}{\partial \dot{\phi}} \dot{\phi}^2 = \dot{\phi}(x).\end{aligned}$$

Therefore the Klein-Gordon Hamiltonian density, using the Eq. (20.2.3), can be written as

$$\begin{aligned}\mathcal{H} &= \pi(x)\dot{\phi}(x) - \mathcal{L} = \dot{\phi}^2 - \frac{1}{2}\partial_\mu \phi \, \partial^\mu \phi + \frac{1}{2}m^2 \phi^2 \\ &= \dot{\phi}^2 - \frac{1}{2}\partial_0 \phi \, \partial^0 \phi - \frac{1}{2}\partial_k \phi \, \partial^k \phi + \frac{1}{2}m^2 \phi^2 \\ &= \frac{1}{2}\pi^2 + \frac{1}{2}(\vec{\nabla}\phi)^2 + \frac{1}{2}m^2 \phi^2.\end{aligned}$$

Chapter 23

The electromagnetic field

23.1 Maxwell's equations

The four Maxwell's equations

$$\begin{cases} \vec{\nabla} \cdot \vec{E} = \rho \\ \vec{\nabla} \cdot \vec{B} = 0 \\ \vec{\nabla} \times \vec{E} = -\frac{\partial \vec{B}}{\partial t} \\ \vec{\nabla} \times \vec{B} = \frac{\partial \vec{E}}{\partial t} + \vec{J} \end{cases},$$

can be written in the covariant form

$$\begin{cases} \partial_\mu F^{\mu\nu} = J^\nu \\ \partial_\mu F^{\nu\rho} + \partial_\nu F^{\rho\mu} + \partial_\rho F^{\mu\nu} = 0 \end{cases},$$

where $F^{\mu\nu}$ is the tensor of the electromagnetic field related to the four-potential

$$F^{\mu\nu} = \partial^\mu A^\nu - \partial^\nu A^\mu$$

and J^μ is the four-current. We recall that the relations between the four-potential

$$A^\mu = (A^0, \vec{A}) = (\phi, \vec{A})$$

which represents the photon field and the electric and magnetic fields are

$$\vec{E} = -\vec{\nabla}\phi - \frac{\partial \vec{A}}{\partial t}, \quad \vec{B} = \vec{\nabla} \times \vec{A},$$

23.2 Gauge invariance

where ϕ is the scalar potential and $\vec{A}$ is the potential vector. The tensor $F^{\mu\nu}$ is manifestly antisymmetric

$$F^{\mu\nu} = -F^{\nu\mu}$$

and its components are

$$F^{\mu\nu} = \begin{pmatrix} 0 & -E^1 & -E^2 & -E^3 \\ E^1 & 0 & -B^3 & B^2 \\ E^2 & B^3 & 0 & -B^1 \\ E^3 & -B^2 & B^1 & 0 \end{pmatrix}.$$

The conservation of the four-current J^μ derives directly from Maxwell's equations, in fact we have

$$\begin{aligned} \partial_\nu J^\nu &= \partial_\mu \partial_\nu F^{\mu\nu} = \frac{1}{2}(\partial_\mu \partial_\nu F^{\mu\nu} + \partial_\nu \partial_\mu F^{\nu\mu}) \\ &= \frac{1}{2}(\partial_\mu \partial_\nu F^{\mu\nu} - \partial_\mu \partial_\nu F^{\mu\nu}) = 0. \end{aligned}$$

23.2 Gauge invariance

We observe that the electromagnetic tensor is invariant under the transformations of the four-potential $A^\mu(x)$ (called gauge transformations)

$$A^\mu(x) \to A'^\mu(x) = A^\mu(x) + \partial^\mu \varphi(x),$$

where $\varphi(x)$ is an arbitrary function. Replacing $A^\mu(x)$ with $A'^\mu(x)$ using the previous equation is related to a particular choice of gauge. Known examples are the Lorentz gauge and the Coulomb gauge. In the Lorentz gauge the four-potential satisfies the relation

$$\partial_\mu A^\mu = 0.$$

In this case the $\varphi(x)$ function must satisfy

$$\Box \varphi(x) = -\partial_\mu A^\mu(x)$$

and therefore this gauge is not univocally determined. In fact we can consider to add any other function such that

$$\Box \varphi = 0.$$

On the other hand, in the Coulomb gauge the relation is

$$\vec{\nabla} \cdot \vec{A} = 0.$$

In the free case, where $J^\mu = 0$, it is possible to use, as a gauge transformation, simultaneously the two relations

$$\vec{\nabla} \cdot \vec{A} = 0, \quad A^0 = \phi = 0.$$

23.3 Maxwell Lagrangian

The free electromagnetic field can be described by the following Lagrangian density (also called Maxwell Lagrangian)

$$\mathcal{L} = -\frac{1}{4} F^{\mu\nu} F_{\mu\nu}, \qquad (23.3.1)$$

where

$$\begin{aligned}
F^{\mu\nu} F_{\mu\nu} &= (\partial^\mu A^\nu - \partial^\nu A^\mu)(\partial_\mu A_\nu - \partial_\nu A_\mu) \\
&= (\partial^\mu A^\nu)(\partial_\mu A_\nu) - (\partial^\nu A^\mu)(\partial_\mu A_\nu) \\
&\quad - (\partial^\mu A^\nu)(\partial_\nu A_\mu) + (\partial^\nu A^\mu)(\partial_\nu A_\mu) \\
&= 2(\partial^\mu A^\nu)(\partial_\mu A_\nu) - 2(\partial^\mu A^\nu)(\partial_\nu A_\mu) \\
&= 2(\partial^\mu A^\nu)(\partial_\mu A_\nu - \partial_\nu A_\mu) = 2(\partial^\mu A^\nu) F_{\mu\nu}.
\end{aligned}$$

In order to use the equations of motion shown in Eq. (20.1.2), which in this case take the form

$$\partial_\mu \frac{\partial \mathcal{L}}{\partial(\partial_\mu A_\nu)} = \frac{\partial \mathcal{L}}{\partial A_\nu},$$

we calculate

$$\frac{\partial \mathcal{L}}{\partial A_\nu} = 0.$$

Knowing that

$$\frac{\partial}{\partial(\partial_\mu A_\nu)} F^{\alpha\beta} = \frac{\partial}{\partial(\partial_\mu A_\nu)}(\partial^\alpha A^\beta - \partial^\beta A^\alpha)$$
$$= \delta^{\alpha\mu}\delta^{\beta\nu} - \delta^{\beta\mu}\delta^{\alpha\nu}$$

and

$$\frac{\partial}{\partial(\partial_\mu A_\nu)} F_{\alpha\beta} = \delta_\alpha^\mu \delta_\beta^\nu - \delta_\beta^\mu \delta_\alpha^\nu.$$

$$\frac{\partial \mathcal{L}}{\partial(\partial_\mu A_\nu)} = -\frac{1}{4}\frac{\partial}{\partial(\partial_\mu A_\nu)}(F^{\alpha\beta}F_{\alpha\beta})$$
$$= -\frac{1}{4}\Big(F_{\alpha\beta}(\delta^{\alpha\mu}\delta^{\beta\nu} - \delta^{\beta\mu}\delta^{\alpha\nu})$$
$$+ F^{\alpha\beta}(\delta_\alpha^\mu \delta_\beta^\nu - \delta_\beta^\mu \delta_\alpha^\nu)\Big)$$
$$= -\frac{1}{4}(F^{\mu\nu} - F^{\nu\mu} + F^{\mu\nu} - F^{\nu\mu}) = -F^{\mu\nu},$$

23.3 Maxwell Lagrangian

we get one of the two sets of Maxwell's equations for the free field (free photon)

$$\partial_\mu F^{\mu\nu} = 0.$$

Chapter 24

The Dirac field

24.1 Dirac equation

The Dirac equation describes, in relativistic quantum mechanics, particles of spin 1/2. The wave function for a given particle is a four-component spinor $\psi(x)$. The Dirac s equation has the form

$$(i\gamma^\mu \partial_\mu - m)\psi(x) = 0, \qquad (24.1.1)$$

where the γ^μ are the four Dirac 4×4 matrices that obey the following anticommutator rule

$$\{\gamma^\mu, \gamma^\nu\} = 2\eta^{\mu\nu} 1, \qquad (24.1.2)$$

where the identity matrix shown in the second member, denoted with 1, (in this case a 4×4 identity matrix) will sometimes be omitted, for simplicity. The Dirac equation is written in terms of the matrices γ^μ and the spinor $\psi(x)$. Writing explicitly the sum on the indices of the components, the Dirac equation can be written has the set of 4 equations ($a = 1,2,3,4$)

$$i \sum_{b=1}^{4} (\gamma^\mu)_{ab} \partial_\mu \psi_b = m \psi_a.$$

24.2 Properties of γ matrices

From the anticommutation rule of Eq. (24.1.2) we immediately obtain the following properties of the Dirac gamma matrices

$$(\gamma^0)^2 = 1, \qquad (\gamma^k)^2 = -1,$$

in fact

$$\{\gamma^0, \gamma^0\} = 2(\gamma^0)^2 = 2\eta^{00} = 2 \cdot 1$$

24.2 Properties of γ matrices

and
$$\{\gamma^k, \gamma^k\} = 2(\gamma^k)^2 = 2\eta^{kk} = -2 \cdot 1.$$

Other properties of the γ matrices are

$$(\gamma^0)^\dagger = \gamma^0, \qquad (\gamma^k)^\dagger = -\gamma^k, \qquad (\gamma^\mu)^\dagger = \gamma^0 \gamma^\mu \gamma^0.$$

The γ matrices have a null trace

$$\text{Tr}(\gamma^\mu) = 0,$$

as every odd product of γ matrices

$$\text{Tr}(\underbrace{\gamma^{\mu_1} \gamma^{\mu_2} \cdots \gamma^{\mu_n}}_{n \text{ odd}}) = 0.$$

The trace of the product of an even number of γ matrices is not identically null and we have, for example

$$\text{Tr}(\gamma^\mu \gamma^\nu) = 4\eta^{\mu\nu}$$

and

$$\text{Tr}(\gamma^\mu \gamma^\nu \gamma^\rho \gamma^\sigma) = 4(\eta^{\mu\nu}\eta^{\rho\sigma} - \eta^{\mu\rho}\eta^{\nu\sigma} + \eta^{\mu\sigma}\eta^{\nu\rho}).$$

Other useful relationships are

$$\gamma_\mu \gamma^\mu = 4, \qquad \gamma_\nu \gamma^\mu \gamma^\nu = -2\gamma^\mu, \qquad \gamma_\rho \gamma^\mu \gamma^\nu \gamma^\rho = 4\eta^{\mu\nu}.$$

The γ matrices have the following form in the representation of Dirac

$$\gamma^0 = \begin{pmatrix} 1 & 0 & 0 & 0 \\ 0 & 1 & 0 & 0 \\ 0 & 0 & -1 & 0 \\ 0 & 0 & 0 & -1 \end{pmatrix}, \qquad \gamma^1 = \begin{pmatrix} 0 & 0 & 0 & 1 \\ 0 & 0 & 1 & 0 \\ 0 & -1 & 0 & 0 \\ -1 & 0 & 0 & 0 \end{pmatrix},$$

$$\gamma^2 = \begin{pmatrix} 0 & 0 & 0 & -i \\ 0 & 0 & i & 0 \\ 0 & i & 0 & 0 \\ -i & 0 & 0 & 0 \end{pmatrix}, \qquad \gamma^3 = \begin{pmatrix} 0 & 0 & 1 & 0 \\ 0 & 0 & 0 & -1 \\ -1 & 0 & 0 & 0 \\ 0 & 1 & 0 & 0 \end{pmatrix}.$$

These can be written, using the three Pauli matrices σ_i and the identity matrix 1 (in this case a 2×2 identity matrix) as follows

24.2 Properties of γ matrices

$$\gamma^0 = \begin{pmatrix} 1 & 0 \\ 0 & -1 \end{pmatrix}, \quad \gamma^j = \begin{pmatrix} 0 & \sigma_i \\ -\sigma_i & 0 \end{pmatrix},$$

with

$$\sigma_1 = \begin{pmatrix} 0 & 1 \\ 1 & 0 \end{pmatrix}, \quad \sigma_2 = \begin{pmatrix} 0 & -i \\ i & 0 \end{pmatrix}, \quad \sigma_3 = \begin{pmatrix} 1 & 0 \\ 0 & -1 \end{pmatrix}.$$

An additional 4×4 matrix, called γ^5, can be defined in this way

$$\gamma^5 \equiv i\gamma^0 \gamma^1 \gamma^2 \gamma^3$$

which satisfies the anticommutation relation

$$\{\gamma^5, \gamma^\mu\} = 0$$

and has the following properties

$$(\gamma^5)^2 = 1, \quad (\gamma^5)^\dagger = \gamma^5$$

and
$$\text{Tr}(\gamma^5) = \text{Tr}(\gamma^5 \gamma^\mu \gamma^\nu) = 0$$

and
$$\text{Tr}(\gamma^5 \gamma^\mu \gamma^\nu \gamma^\rho \gamma^\sigma) = -4i\varepsilon^{\mu\nu\rho\sigma}.$$

We introduce the so-called Feynman "slash notation", that will be used later
$$\slashed{p} \equiv p_\mu \gamma^\mu.$$

We already show some results concerning this notation

$$\text{Tr}(\slashed{p}\slashed{k}) = 4(p \cdot k), \quad \gamma_\mu \slashed{p} \gamma^\mu = -2\slashed{p}, \quad \gamma_\mu \slashed{p} \slashed{k} \gamma^\mu = 4(p \cdot k)$$

and

$$\gamma_\mu \slashed{p}\slashed{k}\slashed{q}\gamma^\mu = -2\slashed{q}\slashed{k}\slashed{p}, \quad \{\slashed{p}, \slashed{k}\} = 2(p \cdot k), \quad \slashed{p}^2 = p^2$$

and

$$\text{Tr}(\slashed{p}\slashed{k}\slashed{q}\slashed{P}) = 4\Big((p \cdot k)(q \cdot P) - (p \cdot q)(k \cdot P) + (p \cdot P)(k \cdot q)\Big).$$

24.3 Dirac Lagrangian

Before writing the Dirac Lagrangian density, let's define the adjoint spinor to ψ as

$$\overline{\psi} \equiv \psi^\dagger \gamma^0,$$

from which, for a non-interacting field,

$$\mathcal{L} = \overline{\psi}(i\gamma^\mu \partial_\mu - m)\psi.$$

24.4 Dirac Hamiltonian

The conjugated momenta to ψ and $\overline{\psi}$ are written as

$$\pi_\psi = \frac{\partial \mathcal{L}}{\partial \dot{\psi}} = i\frac{\partial}{\partial \dot{\psi}}(\overline{\psi}\gamma^\mu \partial_\mu \psi) = i\frac{\partial}{\partial \dot{\psi}}(\overline{\psi}\gamma^0 \dot{\psi}) = i\overline{\psi}\gamma^0,$$

and

$$\pi_{\overline{\psi}} = \frac{\partial \mathcal{L}}{\partial \dot{\overline{\psi}}} = i\frac{\partial}{\partial \dot{\overline{\psi}}}(\overline{\psi}\gamma^\mu \partial_\mu \psi) = 0.$$

Hence the Hamiltonian density

$$\begin{aligned}\mathcal{H} &= \pi_\psi \dot\psi(x) + \pi_{\overline\psi}\dot{\overline\psi} - \mathcal{L} = \pi_\psi \dot\psi(x) - \overline\psi(i\gamma^\mu \partial_\mu - m)\psi \\ &= i\overline\psi\gamma^0\dot\psi(x) - \overline\psi(i\gamma^\mu \partial_\mu - m)\psi = -\overline\psi(i\gamma^k \partial_k - m)\psi \\ &= -i\psi^\dagger\gamma^0\gamma^k\partial_k\psi + \psi^\dagger\gamma^0 m\psi = \psi^\dagger(-i\vec\alpha\cdot\vec\nabla + \beta m)\psi,\end{aligned}$$

where we have defined

$$\alpha^k = \gamma^0\gamma^k, \qquad \beta = \gamma^0$$

and

$$\nabla^k = \partial_k = \frac{\partial}{\partial x^k}.$$

The properties of the α^k and β matrices can be obtained from the properties of the γ matrices and are the following

$$\{\alpha^i, \alpha^j\} = 2\delta^{ij}\mathbf{1}, \quad \{\alpha^i, \beta\} = 0, \quad \beta^2 = 1.$$

To find the Hamiltonian we can proceed by calculating

$$H = \int d^3 x\, \mathcal{H},$$

24.4 Dirac Hamiltonian

or by writing the Dirac equation in Eq. (24.1.1) as follows

$$i\gamma^0 \partial_0 \psi(x) + i\gamma^k \partial_k \psi(x) - m\psi(x) = 0,$$

multiplying both members by γ^0 from the left and remembering the definitions of $\vec{\alpha}$ and β we have

$$i(\gamma^0)^0 \frac{\partial}{\partial t}\psi(x) + i\gamma^0 \gamma^k \partial_k \psi(x) - m\gamma^0 \psi(x) = 0,$$

and

$$i\frac{\partial \psi}{\partial t} = (-i\vec{\alpha} \cdot \vec{\nabla} + \beta m)\psi.$$

Comparing with the Dirac equation written in Hamiltonian form

$$H\psi = i\frac{\partial \psi}{\partial t}$$

we get the Dirac Hamiltonian

$$H = -i\vec{\alpha} \cdot \vec{\nabla} + \beta m.$$

24.5 Free particle solutions

The solutions of the Dirac equation

$$(i\gamma^\mu \partial_\mu - m)\psi(x) = 0,$$

which are also the eigenfunctions of the four-momentum p^μ, can be written in two similar ways, depending on whether the energy value is positive $p^0 > 0$ or negative $p^0 < 0$. Let's introduce, for these two cases,

$$\psi_+(x) = u(p)e^{-ip\cdot x}, \qquad p^0 > 0,$$

and

$$\psi_-(x) = v(p)e^{ip\cdot x}, \qquad p^0 < 0.$$

The factors $u(\vec{p})$ and $v(\vec{p})$ describe the spinorial properties of the particles field. By substituting ψ_+ in the Dirac equation we get

$$\begin{aligned} 0 &= (i\gamma^\mu \partial_\mu - m)\psi_+(x) = (i\gamma^\mu \partial_\mu - m)\left(u(p)e^{-ip\cdot x}\right) \\ &= i\gamma^\mu u(p)(-ip_\mu)e^{-ip\cdot x} - mu(p)e^{-ip\cdot x}, \end{aligned}$$

24.5 Free particle solutions

from which, remembering the Feynman slash notation, $\not{p} \equiv \gamma^\mu p_\mu$,

$$(\not{p} - m)u(p) = 0. \tag{24.5.1}$$

Similarly, for ψ_-, we get

$$\begin{aligned}0 &= (i\gamma^\mu \partial_\mu - m)\psi_+(x) = (i\gamma^\mu \partial_\mu - m)\left(v(p)e^{ip\cdot x}\right) \\ &= i\gamma^\mu v(p)(ip_\mu)e^{ip\cdot x} - mv(p)e^{ip\cdot x},\end{aligned}$$

from which

$$(\not{p} + m)v(p) = 0. \tag{24.5.2}$$

For the adjoint spinors we have

$$\bar{u}(p)(\not{p} - m) = 0,$$

and

$$\bar{v}(p)(\not{p} + m) = 0.$$

In the reference frame at rest ($m \neq 0$), where $p^\mu = (m, \vec{0})$, the Eqs. (24.5.1) and (24.5.2) become

$$(\not{p} - m)u(p_0) = (m\gamma^0 - m)u = (\gamma^0 - 1)u = 0,$$

and

$$(\not{p}+m)v(p_0) = (m\gamma^0+m)v = (\gamma^0+1)v = 0.$$

Recalling the matrix form of the γ^0 matrix in the Dirac representation

$$\gamma^0 = \begin{pmatrix} 1 & 0 & 0 & 0 \\ 0 & 1 & 0 & 0 \\ 0 & 0 & -1 & 0 \\ 0 & 0 & 0 & -1 \end{pmatrix},$$

we have

$$\gamma^0 - 1 = \begin{pmatrix} 0 & 0 & 0 & 0 \\ 0 & 0 & 0 & 0 \\ 0 & 0 & -2 & 0 \\ 0 & 0 & 0 & -2 \end{pmatrix} = -2 \begin{pmatrix} 0 & 0 \\ 0 & 1 \end{pmatrix},$$

$$\gamma^0 + 1 = \begin{pmatrix} 2 & 0 & 0 & 0 \\ 0 & 2 & 0 & 0 \\ 0 & 0 & 0 & 0 \\ 0 & 0 & 0 & 0 \end{pmatrix} = 2 \begin{pmatrix} 1 & 0 \\ 0 & 0 \end{pmatrix},$$

24.5 Free particle solutions

from which the equations

$$\begin{pmatrix} 0 & 0 \\ 0 & 1 \end{pmatrix} u(p_0) = 0,$$

and

$$\begin{pmatrix} 1 & 0 \\ 0 & 0 \end{pmatrix} v(p_0) = 0.$$

There are four independent solutions, two for each equation ($u^{(1)}, u^{(2)}$ and $v^{(1)}, v^{(2)}$), which have the form

$$u^{(1)} = \begin{pmatrix} 1 \\ 0 \\ 0 \\ 0 \end{pmatrix}, \quad u^{(2)} = \begin{pmatrix} 0 \\ 1 \\ 0 \\ 0 \end{pmatrix},$$

and

$$v^{(1)} = \begin{pmatrix} 0 \\ 0 \\ 1 \\ 0 \end{pmatrix}, \quad v^{(2)} = \begin{pmatrix} 0 \\ 0 \\ 0 \\ 1 \end{pmatrix}.$$

These solutions are the states corresponding to the two projections of spin 1/2 for u and v.

To calculate $u(p)$ and $v(p)$ in a generic reference frame, starting from the results obtained in the case of the system at rest, we note firstly that

$$\begin{aligned}(\slashed{p}+m)(\slashed{p}-m) &= (\gamma^\mu p_\mu + m)(\gamma^\nu p_\nu - m) \\ &= \gamma^\mu \gamma^\nu p_\mu p_\nu - m^2 = p^2 - m^2,\end{aligned}$$

is an identity, in fact

$$\slashed{p}^2 = \gamma^\mu \gamma^\nu p_\mu p_\nu = 2\eta^{\mu\nu} p_\mu p_\nu - \gamma^\nu \gamma^\mu p_\mu p_\nu = 2p^2 - \slashed{p}^2.$$

For a real particle $p^2 = m^2$, we have the identity

$$(\slashed{p}+m)(\slashed{p}-m) = 0$$

and we can write, for example,

$$(\slashed{p}-m)\Big((\slashed{p}+m)u^{(r)}(p_0)\Big) = 0,$$

with $r = 1, 2$. From the Eq. (24.5.1) we get

$$u^{(r)}(p) = C_u(\slashed{p}+m)u^{(r)}, \qquad (24.5.3)$$

24.5 Free particle solutions

where C_u is a normalization constant. Similarly

$$v^{(r)}(p) = C_v(-\not{p}+m)v^{(r)}.$$

We define

$$\xi^{(r)} = \begin{pmatrix} \delta_{1r} \\ \delta_{2r} \end{pmatrix} = \begin{cases} \begin{pmatrix} 1 \\ 0 \end{pmatrix} & \text{if } r = 1 \\ \begin{pmatrix} 0 \\ 1 \end{pmatrix} & \text{if } r = 2 \end{cases}$$

and calculate

$$\begin{aligned} u^{(r)}(p) &= C_u(\not{p}+m)u^{(r)} = C_u(\not{p}+m)\begin{pmatrix} \xi^{(r)} \\ 0 \end{pmatrix} \\ &= C_u\left[\begin{pmatrix} 1 & 0 \\ 0 & -1 \end{pmatrix}p^0 - \begin{pmatrix} 0 & \vec{\sigma} \\ -\vec{\sigma} & 0 \end{pmatrix}\cdot\vec{p} \right. \\ &\quad \left. + m\begin{pmatrix} 1 & 0 \\ 0 & 1 \end{pmatrix}\right]\begin{pmatrix} \xi^{(r)} \\ 0 \end{pmatrix} \\ &= C_u\begin{pmatrix} E+m & -\vec{\sigma}\cdot\vec{p} \\ \vec{\sigma}\cdot\vec{p} & -E+m \end{pmatrix}\begin{pmatrix} \xi^{(r)} \\ 0 \end{pmatrix}, \end{aligned}$$

from which

$$u^{(r)}(p) = C_u \begin{pmatrix} (E+m)\xi^{(r)} \\ \vec{\sigma}\cdot\vec{p}\,\xi^{(r)} \end{pmatrix}.$$

The adjoint of $u^{(r)}(p)$ is

$$\begin{aligned}\bar{u}^{(r)}(p) &= u^{(r)\dagger}(p)\gamma^0 = C_u^* \begin{pmatrix} (E+m)\xi^{(r)\dagger} & \xi^{(r)\dagger}\vec{\sigma}\cdot\vec{p} \end{pmatrix} \\ &\cdot \begin{pmatrix} 1 & 0 \\ 0 & -1 \end{pmatrix} = C_u^* \begin{pmatrix} (E+m)\xi^{(r)\dagger} & -\xi^{(r)\dagger}\vec{\sigma}\cdot\vec{p} \end{pmatrix}.\end{aligned}$$

To get the normalization constant we calculate

$$\begin{aligned}\bar{u}^{(r)}(p)u^{(s)}(p) &= |C_u|^2 \begin{pmatrix} (E+m)\xi^{(r)\dagger} & -\xi^{(r)\dagger}\vec{\sigma}\cdot\vec{p} \end{pmatrix} \\ &\cdot \begin{pmatrix} (E+m)\xi^{(s)} \\ \vec{\sigma}\cdot\vec{p}\,\xi^{(s)} \end{pmatrix} = |C_u|^2 \Big((E+m)^2 \\ &\cdot \xi^{(r)\dagger}\xi^{(s)} - \xi^{(r)\dagger}(\vec{\sigma}\cdot\vec{p})^2\xi^{(s)}\Big) \\ &= 2m|C_u|^2(E+m)\delta_{rs},\end{aligned}$$

where we have used

$$\begin{aligned}(\vec{\sigma}\cdot\vec{p})^2 &= \sigma_i p^i \sigma_j p^j = \frac{1}{2}p^i p^j \{\sigma_i,\sigma_j\} = p^i p^j \delta_{ij} \\ &= \vec{p}^{\,2} = E^2 - m^2 = (E+m)(E-m),\end{aligned}$$

24.5 Free particle solutions

and we impose the condition

$$\bar{u}^{(r)}(p)u^{(s)}(p) = 2m\,\delta_{rs}\,.$$

We can therefore write the normalization constant as

$$C_u = \frac{1}{\sqrt{E+m}}$$

and from the Eq. (24.5.3) we have

$$u^{(r)}(p) = \frac{\slashed{p}+m}{\sqrt{E+m}}\,u^{(r)}\,.$$

Chapter 25

Quantum electrodynamics

25.1 Interaction Lagrangian

The free Dirac field equation is invariant under global gauge transformations of the type

$$\psi(x) \to \psi'(x) = e^{i\alpha}\psi(x),$$

with α constant, but it is not invariant under local gauge transformations. In order to be invariant also in the latter case it is necessary to add an (interaction) term to the Lagrangian, as shown to obtain the Eq. (21.2.2). In the case of the electromagnetic interaction we have, from the Eq. (21.2.1) with the connection given in Eq. (21.2.3), the covariant derivative

$$D_\mu \equiv \partial_\mu - ieA_\mu(x).$$

In this way, the Dirac Lagrangian density, invariant under local transformations given by the operator

$$U(x) = e^{-ie\Phi(x)}, \quad \Phi(x) \in \mathbb{R},$$

is the following

$$\mathcal{L} = \overline{\psi}(i\gamma^\mu D_\mu - m)\psi.$$

This Lagrangian density is composed of a part that describes the Dirac free field ($\mathcal{L}_D$) and a part that describes the interaction with the electromagnetic field ($\mathcal{L}_{\text{int}}$). Explicitly we have

$$\begin{aligned}\mathcal{L} &= \mathcal{L}_D + \mathcal{L}_{\text{int}} = \overline{\psi}(i\gamma^\mu \partial_\mu - m)\psi + e\overline{\psi}\gamma^\mu \psi A_\mu \\ &= \overline{\psi}(i\slashed{\partial} - m)\psi + e\overline{\psi}\slashed{A}\psi.\end{aligned}$$

25.1 Interaction Lagrangian

The interaction Lagrangian density between the Dirac field and the electromagnetic one is therefore

$$\mathcal{L}_{\text{int}} = e\overline{\psi}(x)\slashed{A}(x)\psi(x), \qquad (25.1.1)$$

while the full Lagrangian density of the QED, which includes the Dirac free field, the electromagnetic free field ($\mathcal{L}_{\text{em}}$), from Eq. (23.3.1), and their interaction is

$$\begin{aligned}\mathcal{L}_{\text{QED}} &= \mathcal{L}_D + \mathcal{L}_{\text{em}} + \mathcal{L}_{\text{int}} \\ &= \overline{\psi}(i\slashed{\partial} - m)\psi - \frac{1}{4}F^{\mu\nu}F_{\mu\nu} + e\overline{\psi}\slashed{A}\psi.\end{aligned}$$

The interaction Lagrangian density can be written also in the following way, using the four-current J^μ associated with the Dirac field,

$$\mathcal{L}_{\text{int}} = e\overline{\psi}\gamma^\mu\psi A_\mu = eJ^\mu A_\mu.$$

Using the last expression together with the Lagrangian density of the free electromagnetic field, from Eq. (23.3.1), we obtain

$$\mathcal{L}_{\text{em+int}} = -\frac{1}{4}F^{\mu\nu}F_{\mu\nu} + eJ^\mu A_\mu.$$

25.2 Interaction Hamiltonian

From the interaction Lagrangian density shown in Eq. (25.1.1) we can derive the interaction Hamiltonian density, using the Eq. (20.2.3), as follows

$$\begin{aligned}\mathcal{H}_{int} &= \frac{\partial \mathcal{L}_{int}}{\partial \dot{\psi}}\dot{\psi} + \frac{\partial \mathcal{L}_{int}}{\partial \dot{\overline{\psi}}}\dot{\overline{\psi}} + \frac{\partial \mathcal{L}_{int}}{\partial \dot{A}_\mu}\dot{A}_\mu - \mathcal{L}_{int} \\ &= -\mathcal{L}_{int} = -e\overline{\psi}\slashed{A}\psi.\end{aligned} \qquad (25.2.1)$$

25.3 Field operators

In second quantization, defining

$$d^3\tilde{p} \equiv \frac{d^3p}{(2\pi)^3 2E}, \qquad (25.3.1)$$

the field operators associated with $\psi(x)$ and $\overline{\psi}(x)$ (fermions of spin 1/2) can be written as

$$\begin{aligned}\psi(x) &= \psi^{(+)} + \psi^{(-)} = \sum_r \int d^3\tilde{p}\,\Big(u^{(r)}(p)b(p,r)e^{-ip\cdot x} \\ &\quad + v^{(r)}(p)d^\dagger(p,r)e^{ip\cdot x}\Big),\end{aligned}$$

25.3 Field operators

$$\begin{aligned}\overline{\psi}(x) &= \overline{\psi}^{(+)} + \overline{\psi}^{(-)} = \sum_r \int d^3\tilde{p}\left(\overline{v}^{(r)}(p)d(p,r)e^{-ip\cdot x}\right.\\ &\quad \left.+ \overline{u}^{(r)}(p)b^\dagger(p,r)e^{ip\cdot x}\right),\end{aligned}$$

with the only non-vanishing anticommutators

$$\left\{b(p,r),b^\dagger(p',r')\right\} = \left\{d(p,r),d^\dagger(p',r')\right\} = 2E\,\delta_{pp'}\delta_{rr'}.$$

Similarly for the electromagnetic field (photon field), which satisfies

$$A_\mu = A_\mu^\dagger,$$

we have

$$\begin{aligned}A_\mu(x) &= A_\mu^{(+)} + A_\mu^{(-)} = \sum_\lambda \int d^3\tilde{k}\left(\varepsilon_\mu(k,\lambda)a(k,\lambda)e^{-ik\cdot x}\right.\\ &\quad \left.+ \varepsilon_\mu^*(k,\lambda)a^\dagger(k,\lambda)e^{ik\cdot x}\right),\end{aligned}$$

with the only non-vanishing anticommutators

$$\left[a(k),a^\dagger(k')\right] = 2E\,\delta_{kk'}.$$

For a complex scalar field we have

$$\phi(x) = \phi^{(+)} + \phi^{(-)} = \int d^3\tilde{p}\left(a(p)e^{-ip\cdot x} + c^\dagger(p)e^{ip\cdot x}\right),$$

and

$$\phi^\dagger(x) = \phi^{\dagger(+)} + \phi^{\dagger(-)} = \int d^3\tilde{p}\left(c(p)e^{-ip\cdot x} + a^\dagger(p)e^{ip\cdot x}\right),$$

with the only non-vanishing anticommutators

$$\left[a(p), a^\dagger(p')\right] = \left[c(p), c^\dagger(p')\right] = 2E\,\delta_{pp'}.$$

Finally for a real scalar field ($\phi = \phi^\dagger$) we can write

$$\phi(x) = \phi^{(+)} + \phi^{(-)} = \int d^3\tilde{k}\left(a(k)e^{-ik\cdot x} + a^\dagger(k)e^{ik\cdot x}\right),$$

with the only non-vanishing anticommutators

$$\left[a(k), a^\dagger(k')\right] = 2E\,\delta_{kk'}.$$

25.4 The S matrix

We can imagine a physical process of scattering or decay as a transition between two asymptotic states, both eigenstates

25.4 The S matrix

of a free Hamiltonian density $\mathcal{H}_0$, the initial one $|i\rangle$ (corresponding to $t = -\infty$) and the final one $|f\rangle$ (corresponding to $t = +\infty$). This transition is performed by an operator related to the interaction Hamiltonian density $\mathcal{H}_{int}$. The equation of motion in the interaction representation for $|\psi(t)\rangle$ can be written as

$$i\hbar \frac{d}{dt}|\psi(t)\rangle = H_{int}|\psi(t)\rangle,$$

from which

$$|\psi(t)\rangle = |\psi(t_0)\rangle - \frac{i}{\hbar}\int_{t_0}^{t} dt_1 \, H_{int}|\psi(t_1)\rangle.$$

After an iteration we get

$$\begin{aligned}|\psi(t)\rangle &= |\psi(t_0)\rangle - \frac{i}{\hbar}\int_{t_0}^{t} dt_1 \, H_{int}|\psi(t_0)\rangle \\ &+ \frac{(-i)^2}{\hbar^2}\int_{t_0}^{t} dt_1 \, H_{int} \int_{t_0}^{t_1} dt_2 \, H_{int}|\psi(t_2)\rangle,\end{aligned}$$

while, iterating several times, we obtain the complete series

$$\begin{aligned}|\psi(t)\rangle &= \sum_{n=0}^{\infty}\left(\frac{-i}{\hbar}\right)^n \int_{t_0}^{t} dt_1 \, H_{int}(t_1) \\ &\cdots \int_{t_0}^{t_{n-1}} dt_n \, H_{int}(t_n)|\psi(t_0)\rangle.\end{aligned}$$

By introducing the so-called time-ordering we can write

$$|\psi(t)\rangle = \sum_{n=0}^{\infty} \frac{(-i)^n}{n!\hbar^n} \int_{t_0}^{t} dt_1 \cdots \int_{t_0}^{t} dt_n \, \mathrm{T}\Big(H_{\mathrm{int}}(t_1) \cdots H_{\mathrm{int}}(t_n)\Big) |\psi(t_0)\rangle.$$

Using now the relation between Hamiltonian and Hamiltonian density, from Eq. (20.2.2), i.e.

$$H_{\mathrm{int}} = \int d^3x \, \mathcal{H}_{\mathrm{int}},$$

and the fact that, from the Eq. (25.2.1),

$$\int dt \, H_{\mathrm{int}} = -\int dt \, L_{\mathrm{int}} = -\frac{1}{c} \int d^4x \, \mathcal{L}_{\mathrm{int}},$$

we obtain

$$|\psi(t)\rangle = \sum_{n=0}^{\infty} \left(\frac{i}{\hbar c}\right)^n \frac{1}{n!} \prod_{i=1}^{n} \int_{t_0}^{t} d^4x_i \, \mathrm{T}\Big(\mathcal{L}_{\mathrm{int}}(t_1) \cdots \mathcal{L}_{\mathrm{int}}(t_n)\Big) |\psi(t_0)\rangle.$$

We can define the S matrix as

$$|\psi(t)\rangle = \mathrm{U}(t, t_0) |\psi(t_0)\rangle,$$

25.4 The S matrix

with

$$S = \lim_{\substack{t_0 \to -\infty \\ t \to +\infty}} U(t, t_0).$$

The S matrix can therefore be written as a series expansion

$$S = \sum_{n=0}^{\infty} \left(\frac{i}{\hbar c}\right)^n \frac{1}{n!} \prod_{i=1}^{n} \int d^4 x_i \, \mathrm{T}\Big(\mathcal{L}_{\text{int}}(x_1) \cdots \mathcal{L}_{\text{int}}(x_n)\Big),$$

that we can write also as

$$S = S^{(0)} + S^{(1)} + S^{(2)} + \cdots.$$

The first terms are

$$S^{(0)} = 1, \quad S^{(1)} = \frac{i}{\hbar c} \int d^4 x \, \mathrm{T}\Big(\mathcal{L}_{\text{int}}(x)\Big),$$

$$S^{(2)} = -\frac{1}{2\hbar^2 c^2} \int d^4 x \int d^4 y \, \mathrm{T}\Big(\mathcal{L}_{\text{int}}(x)\mathcal{L}_{\text{int}}(y)\Big)$$

and

$$S^{(3)} = -\frac{i}{6\hbar^3 c^3} \int d^4 x \int d^4 y \int d^4 z \, \mathrm{T}\Big(\mathcal{L}_{\text{int}}(x)\mathcal{L}_{\text{int}}(y)\mathcal{L}_{\text{int}}(z)\Big).$$

25. Quantum electrodynamics

Using the so-called normal-ordering the interaction Lagrangian can be written, from Eq. (25.1.1), as

$$\mathcal{L}_{int}(x) = e\,\mathrm{N}\!\left(\overline{\psi}(x)\slashed{A}(x)\psi(x)\right),$$

from which

$$S^{(0)} = 1, \quad S^{(1)} = \frac{ie}{\hbar c}\int d^4x\,\mathrm{T}\!\left(\mathrm{N}\!\left(\overline{\psi}(x)\slashed{A}(x)\psi(x)\right)\right)$$

and

$$\begin{aligned}S^{(2)} = & -\frac{e^2}{2\hbar^2 c^2}\int d^4x \int d^4y\,\mathrm{T}\!\Big(\mathrm{N}\!\left(\overline{\psi}(x)\slashed{A}(x)\psi(x)\right) \\ & \cdot\,\mathrm{N}\!\left(\overline{\psi}(y)\slashed{A}(y)\psi(y)\right)\Big).\end{aligned}$$

Part IV

Condensed Matter Physics

Chapter 26

Introduction

This fourth and final part of the book is an introduction to the condensed matter physics. The main topics are:
- the Drude model;
- the Hall effect;
- the Seebeck effect;
- the thermal conductivity;
- the Sommerfeld model;
- the diffusion;
- the Brownian motion;
- the Fick's laws;
- the Langevin equation;

26. Introduction

- the Fokker-Planck equation;
- the Boltzmann equation;
- the mechanical properties of solids;
- the lattice defects;
- the semiconductors.

Chapter 27

Brownian motion and diffusion

27.1 Introduction

The term Brownian motion refers to the chaotic and disordered motion of particles in a fluid. The name derives from Robert Brown who observed the disordered motion of pollen in water. The motion of the particles is due to the phenomenon of diffusion and depends on various physical quantities, including the temperature.

27.2 Einstein relation

The Einstein-Smoluchowski relationship gives a relation between the diffusion coefficient, the temperature and the mobility of the moving particles. We can imagine two infinitesimal surfaces, parallel and distant Δx. Let n_+ and n_- be the densities of particles after the second surface and before the first one, respectively. The particles current density along the x axis, called J_x, through a space Δx over the time Δt is

$$J_x = \frac{n_- \Delta x - n_+ \Delta x}{\Delta t}.$$

Introducing the mean velocity

$$\frac{\Delta x}{\Delta t} = v_x,$$

since

$$n_+ - n_- \simeq \frac{dn}{dx} \Delta x$$

the particles current density J_x becomes

$$J_x \cong -v_x \Delta x \frac{dn}{dx}. \qquad (27.2.1)$$

27.2 Einstein relation

We define the diffusion coefficient the quantity $\mathcal{D}_x$ such that

$$J_x \cong -\mathcal{D}_x \frac{dn}{dx}, \qquad (27.2.2)$$

therefore

$$\mathcal{D}_x \equiv v_x \Delta x.$$

Using the free mean path ℓ, which also depends on the fluid in which the diffusion occurs, we can assume that

$$\Delta x \cong \frac{1}{3}\ell,$$

in fact, a particle passes from the first surface to the second one traveling a Δx distance. Therefore the Eq. (27.2.1) becomes

$$J_x \cong -\frac{1}{3} v_x \ell \frac{dn}{dx}. \qquad (27.2.3)$$

If we indicate with τ_x the mean time between two consecutive hits of a particle with those of the fluid, we have

$$v_x = \frac{\ell}{3\tau_x}. \qquad (27.2.4)$$

27. Brownian motion and diffusion

We define the mobility μ of a diffusing particle, the following quantity

$$\mu_x \equiv \frac{\tau_x}{m},$$

where m is its mass. Using this definition and the Eq. (27.2.4) in the Eq. (27.2.3), we obtain

$$J_x \simeq -m v_x^2 \mu_x \frac{dn}{dx},$$

that can be written, considering an average in the definition of J_x,

$$J_x \cong -m \langle v_x^2 \rangle \mu_x \frac{dn}{dx}, \qquad (27.2.5)$$

Furthermore, at the equilibrium with temperature T, we can write

$$\frac{1}{2} m \langle v_x^2 \rangle = \frac{1}{2} k_B T,$$

with k_B Boltzmann constant and therefore

$$m \langle v_x^2 \rangle \simeq k_B T,$$

27.3 Fick's laws

substituting in Eq. (27.2.5),

$$J_x \cong -\mu_x k_B T \frac{dn}{dx}.$$

From here, thanks to the Eq. (27.2.2), we can obtain the expression for the diffusion coefficient

$$\mathcal{D}_x = \mu_x k_B T.$$

In three dimensions the previous result can be written as

$$\mathcal{D} = \mu k_B T$$

and the particles current density $\vec{J}$ becomes

$$\vec{J} = -\mathcal{D} \vec{\nabla} n.$$

27.3 Fick's laws

The relation obtained previously, i.e.

$$\vec{J} = -\mathcal{D} \vec{\nabla} n, \qquad (27.3.1)$$

is called Fick's first law. From the number of particles conservation law we can write

$$\vec{\nabla} \cdot \vec{J} = -\frac{\partial n}{\partial t}$$

and, using Eq. (27.3.1), we have

$$\mathcal{D} \nabla^2 n = \frac{\partial n}{\partial t}$$

which represents Fick's second law.

27.4 Random walker

With the term "random walker" we mean the movement along a straight line, from a starting position, where at each step the movement (of length L) can take place to the right or to the left with equal probability of $1/2$. After N steps, the total mean displacement will be zero, i.e.

$$\langle x_N \rangle = 0.$$

The mean square displacement can be calculated as

$$x_N = x_{N-1} \pm L,$$

27.4 Random walker

from which
$$x_N^2 = x_{N-1}^2 + L^2 \pm 2Lx_{N-1}.$$

By averaging we have
$$\langle x_N^2 \rangle = \langle x_{N-1}^2 \rangle + \langle L^2 \rangle \pm 2\langle Lx_{N-1} \rangle = \langle x_{N-1}^2 \rangle + L^2,$$

having used
$$\langle x_{N-1} \rangle = 0.$$

Therefore
$$\langle x_N^2 \rangle = \langle x_{N-1}^2 \rangle + L^2, \quad \forall N \in \mathbb{N}^+.$$

Being $x_0^2 = 0$, at the step $N = 1$ we have
$$\langle x_1^2 \rangle = L^2,$$

moreover
$$\langle x_2^2 \rangle = 2L^2,$$

and
$$\sigma^2 = \langle x_N^2 \rangle = NL^2.$$

The variance grows linearly with the number of steps (i.e. with time) and

$$\sigma = \sqrt{N}L$$

grows linearly with the square root of time.

27.5 Langevin equation

The description of the Brownian motion of particles of mass m can be mathematically described assuming that they are subjected to a viscous friction force of the type[1]

$$-\gamma \dot{x},$$

and to a stochastic force $f(t)$ such that

$$\langle f(t) \rangle = 0$$

and

$$\langle f(t_1)f(t_2) \rangle = \langle f(t_1)^2 \rangle \delta(t_1 - t_2),$$

with no auto-correlation at different times and with

$$\langle x(t)f(t) \rangle = 0. \qquad (27.5.1)$$

---

[1] $\dot{x} = \frac{dx}{dt}$ and $\ddot{x} = \frac{d^2x}{dt^2}$.

27.5 Langevin equation

The equation of motion for the particle of mass m is therefore

$$m\ddot{x}(t) = f(t) - \gamma \dot{x}(t). \qquad (27.5.2)$$

We solve the equation in the variable $\langle x^2(t) \rangle$, instead of $x(t)$. First of all, we can write

$$\frac{d}{dt}x^2 = 2x\dot{x},$$

and

$$\frac{d^2}{dt^2}x^2 = \frac{d}{dt}(2x\dot{x}) = 2\dot{x}^2 + 2x\ddot{x},$$

from which

$$x\dot{x} = \frac{1}{2}\frac{d}{dt}x^2, \qquad (27.5.3)$$

$$x\ddot{x} = \frac{1}{2}\frac{d^2}{dt^2}x^2 - \dot{x}^2. \qquad (27.5.4)$$

Multiplying the Eq. (27.5.2) for x we obtain

$$mx\ddot{x} = xf - \gamma x\dot{x}$$

and, using the Eqs. (27.5.3) and (27.5.4),

$$m\frac{1}{2}\frac{d^2}{dt^2}x^2 - m\dot{x}^2 = xf - \gamma\frac{1}{2}\frac{d}{dt}x^2.$$

Taking the average, we have

$$m\frac{1}{2}\frac{d^2}{dt^2}\langle x^2\rangle - m\langle\dot{x}^2\rangle = \langle xf\rangle - \gamma\frac{1}{2}\frac{d}{dt}\langle x^2\rangle$$

and, from the condition of Eq. (27.5.1), we obtain

$$m\frac{1}{2}\frac{d^2}{dt^2}\langle x^2\rangle - m\langle\dot{x}^2\rangle = -\gamma\frac{1}{2}\frac{d}{dt}\langle x^2\rangle. \qquad (27.5.5)$$

The second first-member term can be written using the relation

$$\frac{1}{2}m\langle\dot{x}^2\rangle = \frac{1}{2}k_B T,$$

with T the temperature and k_B the Boltzmann constant. Therefore the Eq. (27.5.5) becomes

$$\frac{d^2}{dt^2}\langle x^2\rangle + \frac{\gamma}{m}\frac{d}{dt}\langle x^2\rangle - \frac{2k_B}{m}T = 0.$$

27.5 Langevin equation

By defining the quantity

$$z(t) \equiv \frac{d}{dt}\langle x^2(t)\rangle, \qquad (27.5.6)$$

the previous equation becomes

$$\frac{d}{dt}z(t) + \frac{\gamma}{m}z(t) - \frac{2k_B}{m}T = 0.$$

To solve this differential equation we can separate the variables

$$\frac{dz}{\frac{2k_BT}{m} - \frac{\gamma}{m}z(t)} = dt$$

and get the general solution by integrating

$$\int_{z_0}^{z} \frac{1}{z - \frac{2k_BT}{\gamma}} dz = -\frac{\gamma}{m}\int_0^t dt,$$

obtaining

$$z(t) = \frac{2k_BT}{\gamma} + \left(z_0 - \frac{2k_BT}{\gamma}\right)e^{-\frac{\gamma}{m}t}.$$

The term between brackets is a constant and, denoting it with c, we can write the solution

$$z(t) = \frac{2k_B T}{\gamma} + c e^{-\frac{\gamma}{m}t}.$$

The stationary solution can be obtained by taking the limit $t \to \infty$ and it is

$$z = \frac{2k_B T}{\gamma},$$

from which, using the Eq. (27.5.6), we have

$$\frac{d}{dt}\langle x^2 \rangle = \frac{2k_B T}{\gamma}$$

$$\langle x^2(t) \rangle = \frac{2k_B T}{\gamma} t \qquad (27.5.7)$$

and

$$\sigma = \sqrt{\langle x^2(t) \rangle} = \sqrt{\frac{2k_B T}{\gamma}} \sqrt{t},$$

i.e. a result similar to the random walker. In three dimensions and assuming spherical particles of radius R we can use the Stokes's law

$$\gamma = 6\pi\eta R,$$

with η viscosity of the fluid, the Eq. (27.5.7) becomes

$$\langle x^2(t)\rangle = \frac{k_B T}{\pi R \eta}t,$$

where the coefficient

$$\frac{k_B T}{\pi R \eta}$$

is called Einstein coefficient.

27.6 Fokker-Planck equation

The Fokker-Planck equation in one spatial dimension is

$$\frac{\partial f}{\partial t} = -\frac{\partial}{\partial x}\Big(D_1(x,t)f(x,t)\Big) + \frac{\partial^2}{\partial x^2}\Big(D_2(x,t)f(x,t)\Big),$$

where $f(x,t)$ is the probability density function for a particle. The terms $D_1(x,t)$ and $D_2(x,t)$ are called the drift term and the diffusion term, respectively. A generalization of the Fokker-Planck formula is expressed by the Kramers-Moyal expansion

$$\frac{\partial f}{\partial t} = \sum_{k=1}^{\infty} \left(-\frac{\partial^k}{\partial x^k} D_k(x,t) \right) f(x,t).$$

27.7 Boltzmann equation

The Boltzmann transport equation can be considered as a special case of the Fokker-Planck equation. The Boltzmann equation is

$$\frac{\partial f}{\partial t} = -v\frac{\partial f}{\partial x} - \frac{F}{m}\frac{\partial f}{\partial v} + \left(\frac{\partial f}{\partial t}\right)_{coll},$$

where

$$\left(\frac{\partial f}{\partial t}\right)_{coll}$$

is the collision term, v is the velocity, F is the force and m is the mass. From the Liouville theorem for a system without collisions, called $f(x,v,t)$ the classical distribution function related to the particle probability density, we can write

$$\frac{df(x,v,t)}{dt} = 0.$$

27.7 Boltzmann equation

From the definition of total derivative we can write

$$\frac{\partial f}{\partial t} + \frac{\partial f}{\partial x}\frac{\partial x}{\partial t} + \frac{\partial f}{\partial v}\frac{\partial v}{\partial t} = 0,$$

from which

$$\frac{\partial f}{\partial t} = -v\frac{\partial f}{\partial x} - a\frac{\partial f}{\partial v},$$

In the presence of collisions we can add a collisional term

$$\frac{\partial f}{\partial t} = -v\frac{\partial f}{\partial x} - a\frac{\partial f}{\partial v} + \left(\frac{\partial f}{\partial t}\right)_{coll},$$

which is the Boltzmann transport equation. In the so-called relaxation time approximation, the collision term is assumed to have the form

$$\left(\frac{\partial f}{\partial t}\right)_{coll} = -\frac{f - f_{eq}}{\tau},$$

where τ is the mean time between two collisions and f_{eq} is the probability density function at equilibrium. For an electron gas in a metal where each electron is subjected to a force $-eE$ we have, for example,

$$\frac{\partial f}{\partial t} = -v\frac{\partial f}{\partial x} + \frac{eE}{m}\frac{\partial f}{\partial v} - \frac{f - f_{eq}}{\tau}.$$

27. Brownian motion and diffusion

In stationary conditions the previous expression becomes

$$\frac{eE}{m}\frac{\partial f}{\partial v} = \frac{f - f_{eq}}{\tau}$$

and, denoting

$$E_c = \frac{1}{2}mv^2, \quad \frac{\partial f}{\partial v} = \frac{\partial f}{\partial E_c}\frac{\partial E_c}{\partial v} = mv\frac{\partial f}{\partial E_c},$$

we obtain

$$f = f_{eq} + eE\tau v\frac{\partial f}{\partial E_c}.$$

In the case of thermal transport, the Boltzmann equation, in the relaxation time approximation and in stationary conditions, becomes

$$v\frac{\partial f}{\partial x} + \frac{f - f_{eq}}{\tau} = 0.$$

Chapter 28

Drude model

28.1 Introduction

The Drude model assumes that the electrons in a metal behave like a gas and can be considered as spheres of negligible size that do not interact with each other. The only interactions are elastic collision with the walls. There are also collisions with the ions, but the particles are considered free between one collision and the other and the speed after a collision is that compatible with $\langle v^2 \rangle$ typical of the system at a temperature T, or

$$\frac{1}{2}m\langle v^2 \rangle = \frac{3}{2}k_B T.$$

28.2 Electric conductivity

The electrical conductivity σ is expressed by the Ohm's law

$$\vec{j} = \sigma \vec{E}, \qquad (28.2.1)$$

where $\vec{J}$ and $\vec{E}$ are the current density vector and the electric field vector, respectively. An estimate of the electrical conductivity can be made in the Drude model for an electron gas. By indicating with $n(x,y,z)$ the density of the conduction electrons we can write their current density in this way

$$\vec{j} = -en\langle\vec{v}\rangle, \qquad (28.2.2)$$

where e is the modulus of the electron charge and $\langle\vec{v}\rangle$ is the mean velocity. The classic Drude model assumes that the electrons, during a mean time τ, suffer a constant force in time given by

$$\vec{F} = -e\vec{E}, \quad \frac{\partial}{\partial t}\vec{F} = 0,$$

28.2 Electric conductivity

This provides the differential equation of motion for an electron

$$m_e \dot{\vec{v}} = -e\vec{E},$$

where m_e is the electron mass. Integrating between 0 and τ we obtain

$$\vec{v}(t) = \vec{v}(0) - \frac{e}{m_e}\vec{E}\tau$$

and, averaging over all the electrons,

$$\langle \vec{v} \rangle = -\frac{e}{m_e}\vec{E}\tau,$$

in fact

$$\langle \vec{v}(0) \rangle = 0.$$

The current density of conduction electrons, see Eq. (28.2.2), becomes

$$\vec{j} = \frac{e^2 n \tau}{m_e}\vec{E}$$

and, from the Eq. (28.2.1), we obtain the electrical conductivity

$$\sigma = \frac{e^2 n \tau}{m_e}. \qquad (28.2.3)$$

It should be emphasized that the Drude model is incorrect in reality, in fact the electrons, thanks to their wave behavior, are not perturbed by a periodic potential such as that of a perfect crystal lattice (Block's theorem). Electrons can be scattered by phonons or impurities present in the crystal.

28.3 Hall effect

If we apply a magnetic field $\vec{B}$ on a conductor with a current along the x axis of an appropriate reference system the electrons of the current will suffer the Lorentz force given by

$$\vec{F}_L = -\frac{e}{c}\vec{v} \times \vec{B},$$

where $\vec{v}$ is the electron velocity, e is the modulus of the electron charge and c is the speed of light. At equilibrium we can write

$$0 = \frac{d\vec{p}}{dt} = -e\left(\vec{E} + \frac{\vec{p}}{mc} \times \vec{B} - \gamma\vec{p}\right),$$

28.3 Hall effect

with m mass of the electron, from which

$$\begin{cases} -eE_x - p_y\omega - \frac{1}{\tau}p_x = 0 \\ -eE_y + p_x\omega - \frac{1}{\tau}p_y = 0 \end{cases},$$

where

$$\gamma = \frac{1}{\tau}, \quad \omega = \frac{Be}{mc}.$$

Multiplying both members by

$$-\frac{ne\tau}{m}$$

with n the density of the material, we have

$$\begin{cases} \frac{ne^2\tau}{m}E_x + \frac{ne\tau}{m}p_y\omega + \frac{ne}{m}p_x = 0 \\ \frac{ne^2\tau}{m}E_y - \frac{ne\tau}{m}p_x\omega + \frac{ne}{m}p_y = 0 \end{cases}$$

Remembering the Eq. (28.2.3) we obtain

$$\begin{cases} \sigma E_x + \frac{ne\tau}{m}p_y\omega + \frac{ne}{m}p_x = 0 \\ \sigma E_y - \frac{ne\tau}{m}p_x\omega + \frac{ne}{m}p_y = 0 \end{cases}$$

or
$$\begin{cases} \sigma E_x = \tau\omega J_y - J_x \\ \sigma E_y = -\tau\omega J_x + J_y \end{cases}.$$

When the charges are accumulated along the axis orthogonal to the x axis, due to the Hall effect, the term J_y is null, therefore
$$E_y = -\frac{\tau\omega}{\sigma}J_x = -\frac{\tau eB}{mc\sigma}J_x.$$

We can also define the Hall resistance as
$$R_H = \frac{E_y}{BJ_x}$$

and therefore
$$R_H = -\frac{e\tau}{mc\sigma} = -\frac{1}{nec}.$$

This model does not adapt very well to the experimental results and for small magnetic fields there is a larger discrepancy between data and previsions.

28.4 Thermal conductivity

We can connect the thermal energy current density with the gradient of temperature through a constant, called thermal

28.4 Thermal conductivity

conductivity, in this way

$$\vec{J}^T = -\mathcal{K}_x \vec{\nabla} T.$$

By considering only the x axis we can write

$$J_x^T = -\mathcal{K}_x \frac{dT}{dx}.$$

We can the Drude model also to the thermal transport. The thermal energy current density J_x^T can be written as

$$J_x^T = n_e E_T v_x,$$

with E_T the thermal energy and n_e the electron density. This relation becomes

$$J_x^T = \frac{1}{2} n_e v_x \Big(E_T \left(T(x - v_x \tau) \right) - E_T \left(T(x + v_x \tau) \right) \Big),$$

therefore

$$J_x^T = \frac{1}{2} n_e v_x \frac{dE_T}{dT} \frac{dT}{dx} (-2 v_x \tau) = -v_x^2 n_e \tau \frac{dE_T}{dT} \frac{dT}{dx}. \quad (28.4.1)$$

The specific heat per unit volume of the system of N particles in this case is
$$c_V = \frac{N}{V}\frac{dE_T}{dT} = n_e \frac{dE_T}{dT},$$
so that the Eq. (28.4.1) becomes
$$J_x^T = -v_x^2 \tau c_V \frac{dT}{dx},$$
or
$$J_x^T = -\mathcal{K}_x \frac{dT}{dx},$$
with
$$\mathcal{K}_x = v_x^2 \tau c_V,$$
said thermal conductivity coefficient. In three dimensions the equation takes the form
$$\vec{J}^T = -\mathcal{K}_x \vec{\nabla} T.$$

28.5 Seebeck effect

The Seebeck effect concerns the formation of an electric field due to the transport of heat by the electrons. The electric field

28.5 Seebeck effect

is given by
$$\vec{E} = K_S \vec{\nabla} T,$$

with Q Seebeck coefficient. The Drude model gives in this case
$$K_S \simeq -\frac{k_b}{2e},$$

with k_B Boltzmann constant, which is two orders of magnitude smaller than the experimental result.

Chapter 29

Sommerfeld model

29.1 Quantum treatment

This model assumes the hypothesis of free and independent electrons, as in the Drude model, but a quantum treatment is now introduced, using the stationary Schrödinger equation

$$-\frac{\hbar^2}{2m}\nabla^2 \psi(\vec{r}) = \varepsilon \psi(\vec{r}),$$

considering a confinement of the particles in a cube with side length L and volume $V = L^3$. The solutions of the stationary

Schrödinger equation are

$$\psi(\vec{r}) = \frac{1}{\sqrt{V}} e^{i\vec{k}\cdot\vec{r}}, \qquad \varepsilon = \frac{\hbar^2 k^2}{2m}. \qquad (29.1.1)$$

Imposing the so-called Born-Von Karman boundary condition, i.e.

$$\psi(x+L, y, z) = \psi(x, y, z),$$
$$\psi(x, y+L, z) = \psi(x, y, z),$$

and

$$\psi(x, y, z+L) = \psi(x, y, z),$$

we obtain the following quantization relations for the components of the wave vector

$$k_i = \frac{2\pi}{L} n_i, \qquad i = x, y, z.$$

In the ground state at temperature $T = 0$ the electrons, in the wave vectors space, will occupy states with increasing k. The first two electrons will be located in the origin and so on, with a maximum of two electrons for each state, in agreement with the Pauli exclusion principle, being fermions. Doing so, many electrons, of the order of Avogadro's number, will form a sort of sphere in the wave vectors space. This sphere is called the Fermi sphere and the maximum $\vec{k}$ that identifies its "radius" is called the Fermi wave vector and it is denoted with $\vec{k}_F$. Fermi energy is naturally defined as

$$\varepsilon_F := \frac{\hbar^2 k_F^2}{2m}. \qquad (29.1.2)$$

29.2 Internal energy

In the wave vectors space the volume occupied by a vector $\vec{k}$, called $\Delta \vec{k}$, is given by

$$\Delta^3 k = \left(\frac{2\pi}{L}\right)^3 = \frac{8\pi^3}{V}. \qquad (29.2.1)$$

The internal energy U of the electron gas can be formally written as

$$U = 2 \sum_{\vec{k}} \varepsilon(\vec{k}) = \sum_{\vec{k}} \frac{\hbar^2 k^2}{m},$$

29. Sommerfeld model

since there are two electrons for each energy level. The internal energy per unit of volume is

$$u := \frac{U}{V} = \frac{\hbar^2}{mV} \sum_{\vec{k}} k^2.$$

The latter, using the Eq. (29.2.1), can be written also as

$$u = \frac{\hbar^2}{mV} \sum_{\vec{k}} k^2 \frac{\Delta^3 k}{\Delta^3 k} = \frac{\hbar^2}{8\pi^3 m} \sum_{\vec{k}} k^2 \Delta^3 k$$

If the quantity $\Delta^3 k$ is very small, we can switch from a summation to an integral by introducing

$$\Delta^3 k \to d^3 k,$$

obtaining

$$u = \frac{\hbar^2}{8\pi^3 m} \int_{\vec{k}} k^2 d^3 k.$$

Using the spherical coordinates, with

$$d^3 k = k^2 \, dk \, d\Omega = k^2 \sin\theta \, d\theta \, d\phi \, dk,$$

29.2 Internal energy

we have

$$u = \frac{\hbar^2}{8\pi^3 m}\int_0^{2\pi} d\phi \int_0^{\pi} \sin\theta\, d\theta \int_0^{k_F} k^4\, dk = \frac{\hbar^2}{2\pi^2 m}\frac{k_F^5}{5},$$

where k_F is the modulus of the Fermi wave vector. Using the expression of the Fermi energy given in Eq. (29.1.2), the internal energy per unit volume becomes

$$u = \frac{\varepsilon_F k_F^3}{5\pi^2}.$$

The mean energy per electron is

$$E := \frac{U}{N} = \frac{U}{V}\frac{V}{N} = \frac{u}{n},$$

where N is the total number of electrons and n is the electron density defined by

$$n := \frac{N}{V}.$$

We note that the number of states in the wave vectors space is obtained dividing the volume of the Fermi sphere by the volume of a single state, i.e., from Eq. (29.2.1),

$$\frac{4}{3}\pi k_F^3 \frac{V}{8\pi^3} = \frac{V k_F^3}{6\pi^2},$$

since there are two electrons for each state, we obtain

$$N = \frac{V k_F^3}{3\pi^2}$$

and therefore the electron density is

$$n = \frac{N}{V} = \frac{k_F^3}{3\pi^2}. \qquad (29.2.2)$$

Finally the mean energy per electron becomes

$$E = \frac{u}{n} = \frac{\varepsilon_F k_F^3}{5\pi^2} \frac{3\pi^2}{k_F^3} = \frac{3}{5}\varepsilon_F.$$

In case the temperature is different from zero, the internal energy can be written as

$$U = 2\sum_{\vec{k}} \varepsilon(\vec{k}) f(\varepsilon(\vec{k}), T),$$

where $f(\varepsilon, T)$ is the Fermi-Dirac distribution function

$$f(\varepsilon, T) = \frac{1}{1 + e^{\frac{\varepsilon - \mu}{k_B T}}}, \qquad (29.2.3)$$

29.2 Internal energy

with μ is the chemical potential and k_B is the Boltzmann constant. Similarly to the case with $T = 0$ we can write

$$u = \frac{2}{V}\sum_{\vec{k}}\varepsilon(\vec{k})f(\varepsilon(\vec{k}),T)\frac{\Delta^3 k}{\Delta^3 k}$$
$$= \frac{1}{4\pi^3}\sum_{\vec{k}}\varepsilon(\vec{k})f(\varepsilon(\vec{k}),T)\Delta^3 k,$$

switching to the integral notation as previously done

$$u = \frac{1}{4\pi^3}\int_{\vec{k}}\varepsilon(\vec{k})f(\varepsilon(\vec{k}),T)\,d^3k$$
$$= \frac{1}{\pi^2}\int_0^\infty \varepsilon(k)f(\varepsilon(k),T)k^2\,dk,$$

having used, in spherical coordinates,

$$d^3k = k^2\,dk\,d\Omega = k^2\sin\theta\,d\theta\,d\phi\,dk.$$

From the Eq. (29.1.1), we obtain

$$d\varepsilon = \frac{\hbar^2}{m}k\,dk, \qquad k = \sqrt{\frac{2m\varepsilon}{\hbar^2}}$$

and the integral becomes

$$u = \frac{1}{\pi^2} \int_0^\infty \varepsilon f(\varepsilon,T) \frac{m}{\hbar^3}\sqrt{2m\varepsilon}\,d\varepsilon.$$

The density of states is defined as the function $g(\varepsilon)$ such that

$$u = \int_0^\infty \varepsilon f(\varepsilon,T) g(\varepsilon)\,d\varepsilon,$$

therefore

$$g(\varepsilon) := \frac{m}{\pi^2 \hbar^3}\sqrt{2m\varepsilon}. \qquad (29.2.4)$$

From the expression of the Fermi energy of Eq. (29.1.2) we have

$$\frac{m}{\hbar^2} = \frac{k_F^2}{2\varepsilon_F}$$

and the density of states becomes

$$g(\varepsilon) = \frac{m}{\pi^2 \hbar^3}\sqrt{2m\varepsilon} = \frac{1}{\pi^2}\frac{m}{\hbar^2}\sqrt{2\varepsilon \frac{m}{\hbar^2}}$$

$$= \frac{1}{\pi^2}\frac{k_F^3}{2\varepsilon_F}\sqrt{\frac{\varepsilon}{\varepsilon_F}},$$

29.3 Sommerfeld expansion

moreover, recalling the Eq. (29.2.2),

$$g(\varepsilon) = \frac{3}{2\varepsilon_F} \frac{k_F^3}{3\pi^2} \sqrt{\frac{\varepsilon}{\varepsilon_F}} = \frac{3}{2} \frac{n}{\varepsilon_F} \sqrt{\frac{\varepsilon}{\varepsilon_F}}$$

and

$$g(\varepsilon_F) = \frac{3}{2} \frac{n}{\varepsilon_F}. \qquad (29.2.5)$$

29.3 Sommerfeld expansion

In the case of

$$\varepsilon_F \gg k_B T,$$

we have

$$\varepsilon_F \approx \mu$$

and we can write, for a function $A(\varepsilon)$, the following series expansion, known as Sommerfeld expansion,

$$\int_0^\infty A(\varepsilon) f(\varepsilon, T) d\varepsilon \simeq \int_0^\mu A(\varepsilon) d\varepsilon + \frac{\pi^2}{6} (k_B T)^2 \frac{dA}{d\varepsilon}\bigg|_\mu + \mathcal{O}\left(\frac{k_B T}{\mu}\right)^4,$$

where $f(\varepsilon, T)$ is the Fermi-Dirac distribution function shown in Eq. (29.2.3). Truncating on the first order and knowing that $\varepsilon_F \approx \mu$, we can write

$$\int_0^\infty A(\varepsilon)f(\varepsilon,T)\,d\varepsilon \simeq \int_0^\mu A(\varepsilon)\,d\varepsilon + \frac{\pi^2}{6}(k_BT)^2\frac{dA}{d\varepsilon}\bigg|_\mu$$

$$= \int_0^{\varepsilon_F} A(\varepsilon)\,d\varepsilon + \int_{\varepsilon_F}^\mu A(\varepsilon)\,d\varepsilon$$

$$+ \frac{\pi^2}{6}(k_BT)^2 A'(\mu)$$

$$\simeq \int_0^{\varepsilon_F} A(\varepsilon)\,d\varepsilon + (\mu - \varepsilon_F)A(\varepsilon_F)$$

$$+ \frac{\pi^2}{6}(k_BT)^2 A'(\varepsilon_F). \qquad (29.3.1)$$

The integrals to be calculated are

$$n = \int_0^\infty f(\varepsilon,T)g(\varepsilon)\,d\varepsilon$$

and

$$u = \int_0^\infty \varepsilon f(\varepsilon,T)g(\varepsilon)\,d\varepsilon, \qquad (29.3.2)$$

29.3 Sommerfeld expansion

For the first one, assuming $A(\varepsilon) = g(\varepsilon)$ in the Eq. (29.3.1), we obtain

$$\begin{aligned} n &= \int_0^\infty f(\varepsilon,T) g(\varepsilon) \, d\varepsilon \\ &= \int_0^{\varepsilon_F} g(\varepsilon) \, d\varepsilon + (\mu - \varepsilon_F) g(\varepsilon_F) \\ &+ \frac{\pi^2}{6} (k_B T)^2 g'(\varepsilon_F), \end{aligned}$$

the derivative $g'(\varepsilon)$ is

$$g'(\varepsilon) = \frac{3}{2} \frac{n}{\varepsilon_F} \sqrt{\frac{1}{\varepsilon_F}} \frac{1}{2\varepsilon}$$

and we have

$$g'(\varepsilon_F) = \frac{3}{4} \frac{n}{\varepsilon_F^2} \sqrt{\frac{1}{\varepsilon_F}},$$

moreover

$$\int_0^{\varepsilon_F} g(\varepsilon) \, d\varepsilon = n,$$

in fact the density is the same, therefore

$$(\mu - \varepsilon_F) g(\varepsilon_F) + \frac{\pi^2}{6} (k_B T)^2 g'(\varepsilon_F) = 0 \qquad (29.3.3)$$

and
$$\mu = \varepsilon_F - \frac{\pi^2}{6}(k_BT)^2\frac{g'(\varepsilon_F)}{g(\varepsilon_F)}.$$

Using the Eq. (29.2.5) we can write
$$\frac{g'(\varepsilon_F)}{g(\varepsilon_F)} = \frac{1}{2\varepsilon_F}$$

from which
$$\mu = \varepsilon_F - \frac{\pi^2}{12\varepsilon_F}(k_BT)^2,$$

or also
$$\mu = \varepsilon_F\left(1 - \frac{1}{3}\left(\frac{\pi k_BT}{2\varepsilon_F}\right)^2\right).$$

For the second integral, see Eq. (29.3.2), assuming $A(\varepsilon) = \varepsilon g(\varepsilon)$ in the Eq. (29.3.1), we obtain

$$\begin{aligned}u &= \int_0^\infty \varepsilon f(\varepsilon,T)g(\varepsilon)\,d\varepsilon \\ &= \int_0^{\varepsilon_F} \varepsilon g(\varepsilon)\,d\varepsilon + (\mu - \varepsilon_F)\varepsilon_F g(\varepsilon_F) \\ &\quad + \frac{\pi^2}{6}(k_BT)^2\left(g'(\varepsilon_F)\varepsilon_F + g(\varepsilon_F)\right) \\ &= u_0 + \frac{\pi^2}{6}(k_BT)^2 g(\varepsilon_F) \\ &\quad + \varepsilon_F\left((\mu - \varepsilon_F)g(\varepsilon_F) + \frac{\pi^2}{6}(k_BT)^2 g'(\varepsilon_F)\right).\end{aligned}$$

29.3 Sommerfeld expansion

Using the Eq. (29.3.3) we see that the last term in parentheses is vanishing, therefore

$$u(T) = u_0 + \frac{\pi^2}{6}(k_B T)^2 g(\varepsilon_F)$$

and, using the Eq. (29.2.5), we obtain the internal energy per unit volume

$$u(T) = u_0 + \frac{\pi^2 n}{4\varepsilon_F}(k_B T)^2.$$

The specific heat per volume unit is

$$c_V = \frac{\partial u}{\partial T} = \frac{\pi^2 n}{2\varepsilon_F}(k_B)^2 T = \frac{\pi^2}{2}\frac{k_B T}{\varepsilon_F} n K_B.$$

The contribution of electrons to the specific heat of a metal (purely quantum contribution) is linear in temperature. Experimentally it occurs that for a metal

$$c_V = AT + BT^3,$$

where A and B are coefficients. The specific heat carried by the electron gas is dominant at low temperatures, while the remaining heat, given by the lattice vibrations (phonons), dominates at room temperature or higher.

Chapter 30

Mechanical properties of solids

30.1 Introduction

In general, a solid subjected to a stress can be deformed. These deformations can be permanent or temporary, depending on the intensity of the stress. Elastic deformations are those for which the solid returns to its original size when the stress ceases and generally follow the Hooke's law. In a material subjected to a stress there is firstly an elastic phase, then a plastic phase in which there is no return to the original dimensions when the force ceases and, finally, there is the breaking of the material, if the stress continues.

30.2 Young's modulus

In the elastic regime a body that undergoes compression or traction tends to shorten or lengthen. Young's modulus takes into account the relation between the applied stress σ, (dimensionally the ratio between a force and a surface) and the following deformation $\frac{\Delta l}{l}$, called strain. In formulas

$$E = \frac{\sigma}{\Delta l/l}.$$

30.3 Poisson's ratio

If a body is subjected to traction or compression, in addition to a change in length along the direction of the stress, a change in transverse dimensions is also observed. For example, for a cylinder with a base radius R and height L we can write

$$\frac{\Delta R}{R} = -\nu \frac{\Delta L}{L},$$

with ν is called Poisson's ratio.

Chapter 31

Lattice defects

31.1 Introduction

Defects play a fundamental role in crystals. They can be of three types: point, line or surface. If a non-constant force is applied, the defects can acquire a mobility and these can also interfere with each other. The material can acquire more rigidity and the plastic regime becomes closer, with an increasing of fragility. Some defects can be reduced by heating the material.

31.2 Point defects

Point defects are divided into vacancy and interstitial. The former indicate the lack of an ion in a lattice site, while the latter indicate the presence of an ion in excess in an interstitial position. The number of vacancy defects can be estimated in a crystal at the thermodynamic equilibrium. In fact the Gibbs free energy can be written as

$$G = U - TS + PV,$$

we denote with $(N+n)$ the total number of lattice points, with N the number of ions actually present and with n the number of vacancy defects, with $n \ll N$. There are

$$\frac{(N+n)!}{n!N!}$$

ways to choose the n defects in $(N+n)$ lattice sites, therefore we have to add to the entropy the term

$$S_1 = k_B \ln\left(\frac{(N+n)!}{n!N!}\right).$$

31.2 Point defects

The Gibbs free energy becomes

$$G(n) = (U - TS) - Tk_B \ln\left(\frac{(N+n)!}{n!N!}\right) + P(N+n)V_0,$$

where V_0 is the volume occupied by an ion of the lattice site, moreover

$$\frac{\partial G(n)}{\partial n} = \frac{\partial (U-TS)}{\partial n} - Tk_B \frac{\partial}{\partial n} \ln\left(\frac{(N+n)!}{n!N!}\right) + PV_0.$$

Using the Stirling formula for large n

$$n! \sim \sqrt{2\pi n}\, \frac{n^n}{e^n},$$

or

$$\ln n! \sim (\ln n - 1)n,$$

we obtain

$$\frac{\partial}{\partial n} \ln\left(\frac{(N+n)!}{n!N!}\right) \simeq \ln\left(\frac{N}{n}\right),$$

and

$$\frac{\partial G(n)}{\partial n} \simeq \frac{\partial (U-TS)}{\partial n} + PV_0 - Tk_B \ln\left(\frac{N}{n}\right).$$

Being $n \ll N$ we can write

$$\frac{\partial(U-TS)}{\partial n} \simeq \left(\frac{\partial(U-TS)}{\partial n}\right)\bigg|_{n=0} = \varepsilon$$

from which

$$\frac{\partial G(n)}{\partial n} \simeq \varepsilon + PV_0 - Tk_B \ln\left(\frac{N}{n}\right).$$

Assuming

$$\frac{\partial G(n)}{\partial n} = 0,$$

we have the solution

$$n = Ne^{-\frac{\varepsilon + PV_0}{k_B T}}$$

which minimizes the Gibbs free energy. The term ε is a good approximation of the energy needed to remove an ion. Furthermore, at atmospheric pressure the term PV_0 is negligible, therefore

$$n \simeq Ne^{-\beta \varepsilon},$$

with

$$\beta = \frac{1}{k_B T}.$$

If the defects were of two types (the number of the first ones indicated with n_+ and the number of the second ones indicated with n_-), due to the neutrality of charge (the total charge of the defects is zero), we obtain the following formula, similar to the case of only one type of defects,

$$n_+ n_- = N_+ N_- e^{-\beta(\varepsilon_+ + \varepsilon_-)},$$

from which also

$$n_+ = n_- = \sqrt{N_+ N_-} e^{-\beta(\varepsilon_+ + \varepsilon_-)/2}.$$

If there is an equal number of positive and negative vacancy defects they are called Schottky defects, on the contrary, if there is the same number of vacancy and interstitials of the same ion they are called Frenkel defects.

31.3 Color centers

To balance the missing charge of a negative ion vacancy, an electron can move near the point of the defect. This electron can be considered as an electron bound by a positively charged center and therefore it will present a certain spectrum of energy levels. The excitations between these levels are responsible for the color that the crystal assumes. An electron bounded to a positive vacancy gives a color center called F-center. There are also the M and R-centers where there are, respectively, two vacancies that bind two electrons and three vacancies that bind three electrons. The most abundant are the F-centers.

31.4 Dislocations

The most common line defects are called dislocations, they are largely responsible for the plastic deformation of the solids. In fact, a model that is based on a perfect crystal does not produce adequate estimates for plastic deformation, it is necessary to add the defects to the crystalline structure to overcome this problem. The most important dislocations are the edge and the screw ones. By heating the material you can have a displacement of the dislocations. When a moving dislocation encounters a point defect, the motion of the dislocation can be interrupted and the material becomes harder and more fragile (decrease of the plastic regime length).

Chapter 32

Semiconductors

32.1 Intrinsic semiconductor

Semiconductors are materials characterized by a relatively small band gap between the valence band and the conduction band. Due to the effect of the temperature, an electron can acquire sufficient energy (of the order of $k_B T$) and pass into the conduction band. The intrinsic semiconductors are pure semiconductors, made up of a single element. Let n_c and p_v be the densities of charge carriers in the conduction band and valence band, respectively. Given the density of the states

$g(\varepsilon)$ and the distribution function

$$f(\varepsilon) = \frac{1}{e^{\beta(\varepsilon-\mu)} + 1},$$

we can write

$$n_c = \int_{E_c}^{+\infty} g_c(\varepsilon) \frac{1}{e^{\beta(\varepsilon-\mu)} + 1} d\varepsilon,$$

and

$$p_v = \int_{-\infty}^{E_v} g_v(\varepsilon) \left(1 - \frac{1}{e^{\beta(\varepsilon-\mu)} + 1}\right) d\varepsilon,$$

from which

$$p_v = \int_{-\infty}^{E_v} g_v(\varepsilon) \frac{1}{e^{\beta(\mu-\varepsilon)} + 1} d\varepsilon.$$

Under the conditions

$$\beta(\varepsilon_c - \mu) \gg 1, \quad \beta(\mu - \varepsilon_v) \ll 1,$$

the previous expressions become

$$n_c = N_c e^{-\beta(\varepsilon_c - \mu)},$$

32.1 Intrinsic semiconductor

and
$$p_v = P_v e^{-\beta(\mu - \varepsilon_v)},$$

with
$$N_c = \int_{E_c}^{+\infty} g_c(\varepsilon) e^{-\beta(\varepsilon - \varepsilon_c)} d\varepsilon,$$

and
$$P_v = \int_{-\infty}^{E_v} g_v(\varepsilon) e^{-\beta(\varepsilon_v - \varepsilon)} d\varepsilon.$$

We can write

$$n_c p_v = N_c P_v e^{-\beta(\varepsilon_c - \varepsilon_v)} = N_c P_v e^{-\beta E_g},$$

called the law of mass action. For the conservation of the charge we have, in the intrinsic case, $n_c = p_v$, therefore

$$n_c = p_v = \sqrt{N_c P_v} e^{-\beta E_g / 2}.$$

Typically the density of states is

$$g_c(\varepsilon) = \frac{1}{\hbar^3 \pi^2} \sqrt{2|\varepsilon - \varepsilon_c| m_{c_{eff}}^3},$$

and
$$g_v(\varepsilon) = \frac{1}{\hbar^3 \pi^2}\sqrt{2|\varepsilon - \varepsilon_c|m_{v_{eff}}^3},$$

from which we have
$$N_c = \frac{1}{4}\left(\frac{2m_{c_{eff}}k_B T}{\pi \hbar^2}\right)^{3/2},$$

and
$$P_v = \frac{1}{4}\left(\frac{2m_{v_{eff}}k_B T}{\pi \hbar^2}\right)^{3/2}.$$

32.2 Extrinsic semiconductor

In the case of extrinsic semiconductors there is an addition of charge carriers with the doping of the material, by adding impurities. We have

$$n_c - p_v = \Delta n \neq 0$$

and we can define
$$n_i = \sqrt{n_c p_v}.$$

Let N_d be the density of donors and N_a the density of acceptors. The average occupation for a ε_d level added by donor

32.2 Extrinsic semiconductor

impurities is

$$\langle n_d \rangle = \frac{\sum_j N_j e^{-\beta(E_j - \mu N_j)}}{\sum_j e^{-\beta(E_j - \mu N_j)}} \simeq \frac{1}{\frac{1}{2} e^{E_d - \mu} + 1},$$

while for acceptors

$$\langle n_a \rangle \simeq \frac{1}{\frac{1}{2} e^{\mu - \varepsilon_a} + 1}.$$

So the density of donor charge carriers is

$$n_d = \frac{N_d}{\frac{1}{2} e^{E_d - \mu} + 1},$$

while that of the acceptor carriers is

$$p_v = \frac{N_a}{\frac{1}{2} e^{\mu - \varepsilon_a} + 1}.$$

Printed in France by Amazon
Brétigny-sur-Orge, FR